HISTORY OF BINARY

AND OTHER NONDECIMAL

NUMERATION

HISTORY OF BINARY

AND OTHER NONDECIMAL

NUMERATION

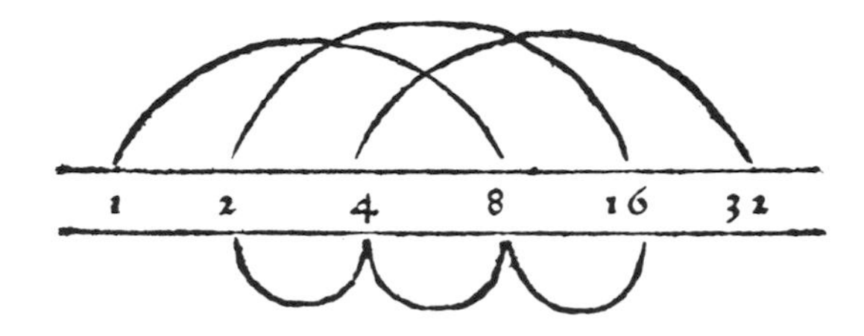

BY ANTON GLASER

Professor of Mathematics, Pennsylvania State University

TOMASH PUBLISHERS

Printed in the United States of America

Library of Congress Cataloging in Publication Data

Glaser, Anton, 1924-
History of binary and other nondecimal numeration.

Based on the author's thesis (Ph.D.—Temple University), presented under the title: History of modern numeration systems.
Bibliography: p. 193
Includes index.
1. Numeration—History. I. Title.
QA141.2.G55 1981 513'.5 81-51176
ISBN 0-938228-00-5 AACR2

To My Wife, Ruth

ACKNOWLEDGMENTS

THIS BOOK is based on the author's doctoral dissertation, *History of Modern Numeration Systems*, written under the guidance of Morton Alpren, Sara A. Rhue, and Leon Steinberg of Temple University in Philadelphia, Pa.

Extensive help was received from the libraries of the Academy of the New Church (Bryn Athyn, Pa.), the American Philosophical Society, Pennsylvania State University, Temple University, the University of Michigan, and the University of Pennsylvania. The photograph of Figure 7 was made available by the New York Public Library; the library of the University of Pennsylvania is the source of the photographs in Figures 2 and 6.

The author is indebted to Harold Hanes, Joseph E. Hofmann, Donald E. Knuth, and Brian J. Winkel, who were kind enough to communicate their comments about the strengths and weaknesses of the original edition. The present revised edition is the better for it. A special thanks is also owed to John Wagner for his careful editorial work and to Adele Clark for her thorough preparation of the Index.

CONTENTS

PAGE

FOREWORD xiii

I. INTRODUCTION 3

II. BEFORE LEIBNIZ 11
Thomas Hariot, Simon Stevin, Francis Bacon, Claude-Gaspar Bachet, Blaise Pascal, Juan Caramuel y Lobkowitz, Erhard Weigel, Joshua Jordaine's *Duodecimal Arithmetick* (1687), Weigel's *Philosophia Mathematica* (1693)

III. LEIBNIZ 31
Letter to the Duke (1697), Two letters to Johann Christian Schulenburg (1698), Correspondence (1701) with Johann (Jean) Bernoulli, Explication (1703), "Nouvelle Arithmétique" by Fontenelle (1703), Thomas Fantet de Lagny, Correspondence (1704–5) with Jacques Bernoulli, Mahnke's Report on Leibniz Manuscripts, Some Reactions to Leibniz's Binary Interpretation of the Figures of Fohy

IV. THE REST OF THE 1700s 53
Some Early Scholarly Sequels to Leibniz's "Explication", Charles XII and his Science Advisor Swedenborg, Euler's and Beguelin's Use of Dyadics, Miscellaneous Writers of the Mid 1700s, Felkel's 1785 Paper on the Periods of β-adic Fractions, Duodecimal versus a Decimal Metric System, Lagrange, Laplace, Lamarque, and Legendre

V. THE NINETEENTH CENTURY 79
Gauss's *Disquisitiones Arithmeticae*, Ozanam's *Recreations in Mathematics*, Peter Barlow, Charles Hutton, John Quincy Adams, John Leslie, Heinrich Wilhelm Stein, André-Marie Ampère, Augustin-Louis Cauchy, D. Vicente Pujals, Augustus DeMorgan, Sir Isaac Pitman, Aimé Mariage, John William Nystrom, E. Stahlberger, A. Sonnenschein, J. W. L. Glaisher, Hermann Hankel, Moritz Cantor, Felix Müller,

P. A. MacMahon, E. Collignon, Vittorio Grünwald, Benjamin Peirce, G. Bellavitis, Lloyd Tanner, Franz Hocevar, Charles Berdellé, Oskar Simony, Edouard Lucas, Hermann Scheffler, Georg Cantor, F. J. Studnicka, Robert M. Pierce, Giuseppe Peano, Alfred B. Taylor, William Woolsey Johnson, E. Gelin, T. N. Thiele, Herbert Spencer, Edward Brooks, School Texts, 1875–1899

VI. THE TWENTIETH CENTURY UP TO THE COMPUTER AGE 115
Moritz Cantor, Binary Numeration Applied to the Game of NIM, The Role of Numeration Systems in Books on Recreational Mathematics and Books on the Theory of Numbers, Miscellaneous Publications on Numeration Systems, Duodecimal Advocates, 1900–1946, Nondecimal Numeration in Textbooks, 1900–1946

VII. APPLICATIONS TO COMPUTERS 133
Number Representation in the ENIAC, The Burks-Goldstine-Von Neumann Report, Some Pre-ENIAC Milestones Toward the Age of Electronic Computers, Four-bit Decimal Codes, Decimal Codes that are Longer than Four Bits, Binary versus Decimal, Alphanumeric Codes, Reflected or Gray Codes

VIII. CONTEMPORARY LITERATURE 159
Some Noteworthy Articles, Amount and Extent of Coverage of Numeration Systems in College Textbooks for Future Elementary School Teachers, Nondecimal Numeration in the SMSG Seventh Grade Material, Contemporary References for Teachers

IX. SUMMARY 169
Chronology, Which Base is Best?

APPENDICES 177

BIBLIOGRAPHY 193

INDEX 211

LIST OF FIGURES

PAGE

1. Two Versions of Leibniz's Design of the Binary Medallion. 32
2. Facsimile of Title Page of Rudolf Nolte's 1734 booklet, in which Leibniz's 1697 letter to the Duke appears as well as the 1703 "Explication." 34
3. Table from Leibniz's May 17, 1698 letter to Johann Christian Schulenberg. 36
4. Table of Numbers from Leibniz's "Explication." 38
5. Examples of Arithmetic Operations Shown in Leibniz's "Explication." 40
6. Facsimile of Page 228 of Carus's 1896 Article in *The Monist*. 50
7. Facsimile of the Title Page of the 1718 Post-doctoral Thesis Written by Johann Bernard Wiedeburg of Jena. 56
8. The Subdivisions of Mathematics According to Ampère's Classification Scheme of 1838. 86
9. The Five Müller Cards of 1876. 86
10. Benjamin Peirce's 'Improved' Binary Notation Compared with Leibniz's. 98
11. Numeric Values (in Binary, Hexadecimal, and Decimal) Assigned to 25 Letters of the Alphabet and the Ten Decimal Digits by Peano. 106
12. Geometric Model for 3-bit Reflected Codes. 154

LIST OF TABLES

PAGE

1. Divisibility Theorems Referred to by Pascal in his "De Numeris Multiplicibus" of 1665. 19

2. Selected 4-bit Decimal Codes and their Characteristics. 142

3. Selected Decimal Codes whose Length is Greater than Four Bits. 144

4. Comparison of Output from a Decimal and a Binary Computer. 150

5. Selected Alphanumeric Codes. 152

6. N-bit Reflected Codes for N = 1 to N = 5 that Suggest a Method of Generating Longer Reflected Codes. 156

7. Content Analysis of Twelve College Textbooks for Future Elementary Teachers with Respect to Numeration Systems. 164

APPENDICES

PAGE

TABLES DEALING WITH WHOLE NUMBERS IN β-ADIC NOTATION

A. Binary Representation of Numbers Compared to More Common Ones 177

B. Comparison of Bases Two to Sixteen 178

C. Binary Representation Contrasted with BCD for Selected Numbers 179

D. Comparison of Hexadecimal and Decimal Numerals 180

E. Selected Numbers Represented in Various Codes 181

TABLES DEALING WITH FRACTIONS IN IN β-ADIC NOTATION

F. Decimal and Binary Equivalents of Fractions of the Type n/64 from n = 0 to n = 63 182

G. Binary and Other Equivalents and Approximations of Fractions of the Type n/10 from n = 0 to n = 9 183

H. Binary and Other Equivalents and Approximations of Unit Fractions from 1/2 to 1/16 184

I. The Number 1/7 Expressed in Bases Two to Sixteen 185

FONTENELLE'S ARTICLE (1703) 187

FOREWORD

WE LIVE in an age of growing interest in the history of science. The reasons for this are many and often obscure, but surely our increasingly complex technology and the questions it has raised about itself are among the most important. Laymen, even more than specialists, are interested in the history of science upon which that technology is based.

Of all the branches of technology, the computer looms largest in the average person's view. The rise of the computer in a very short time has been spectacular, and no end is in sight. For this reason, both the history of computing and the history of the binary system in which computers work are of great interest. The binary system is now constantly taught in schools to illustrate, by contrast, the standard decimal system.

Binary notation is involved in a number of different aspects of computers. Most computers use both computer binary arithmetic and binary logic extensively. Although often hidden slightly, the arithmetic of finite fields of characteristic two—simply binary arithmetic without the carry—is also used in the format error checking codes of computers.

This book is the first carefully researched history of the binary literature. Not only does it cover binary arithmetic, it also includes a number of related topics such as arithmetic in the bases 4, 8, 10, 12, 20, and 60. The base 12, the duodecimal system, is of particular interest because of its many passionate devotees who have long wanted all of our number system converted to it. Thus this book makes a significant contribution to our understanding of the complex world in which we live.

RICHARD W. HAMMING

HISTORY OF BINARY

AND OTHER NONDECIMAL

NUMERATION

·I·

INTRODUCTION

DURING THE last two decades, nondecimal numeration, such as binary and duodecimal, has become a popular topic in the mathematics programs of many primary and secondary schools in the United States. The purpose of this study is to trace the development of nondecimal numeration from the sixteenth century to the present.

The mathematician W. W. Sawyer, who was educated in England and who has been active in teacher education in the United States, wrote in 1964:

> The idea that the writing of numbers need not be based on 10 is not a new one. How old it is I do not know. It was certainly current in the 1880s; Hall and Knight's *School Algebra* and *Higher Algebra* both contain chapters on this subject. In the 1940s Leicester Teacher Training College required their students to work out how the multiplication tables and other parts of arithmetic would have looked if we had possessed 8 fingers instead of 10. Recently the study of 'bases other than 10' became very popular in the United States. The use of the binary scale in electronic computers helped to bring it back into fashion.[1]

The first large-scale electronic digital computer making use of binary numeration appeared in the late 1940s. By 1953 Phillip S. Jones could report:

> The binary system as a special case of the generalized problem of scales of notation has had a sudden resurgence of popularity. This is largely due to its use in modern high-speed electronic calculators and in the new developments in the theory of "information" and "communication." However, this new utility of the binary system arrived at the same time that an even greater emphasis was being placed on 'meaning' and

> 'understanding' in the teaching of mathematics. In arithmetic (and algebra) many teachers have felt that understanding of our number system was enhanced, and in some cases first achieved, through a study of numbers written to some base other than ten.[2]

In 1957 the Commission on Mathematics recommended that one of the goals of seventh and eighth grade arithmetic should be "The understanding of a place system of numeration with special references to the decimal system and the study of other bases, particularly the binary system."[3] This recommendation was seconded by the School Mathematics Study Group (SMSG) in 1959 when it observed that

> since in using a new base the student is forced to look at the reasons for 'carrying' and the other mechanical operations in a new light, he should gain deeper insight into the decimal system. A certain amount of computation in other systems is necessary to fix these ideas but such computation should not be regarded as an end in itself.[4]

In 1961 the National Council of Teachers of Mathematics added its weight to these recommendations and also emphasized the pedagogical rather than the utilitarian (applications to computers) benefits of nondecimal numeration.

The possible utilitarian benefits of a study of nondecimal numeration are not totally absent even for a social scientist who expects to make only occasional use of a computer through a FORTRAN type language. Daniel D. McCracken initially encouraged the contrary view in his 1961 booklet, *A Guide to* FORTRAN *Programming*: "We, however, are not required to study binary numbers, since FORTRAN handles conversions between binary and decimal."[5] But by 1965 McCracken relented and discussed some of the unpleasant surprises that may be in store for the FORTRAN user ignorant of the binary system.[6]

It is not surprising that the double revolution in school mathematics and computers caught most American teachers without a knowledge of nondecimal numeration. Having never heard of this topic during their own school days, they now had to master it as adults. After all, since the sixteenth century, the decimal (i.e., Hindu-Arabic) system had been *the* system for representing numbers in Western civilization, so much so, that the word 'number' had come to mean 'decimal representation' as well. The residual use of Roman numerals hardly seemed relevant. Indeed, as E. T. Bell reminded us, "neither the Hindu numerals nor any others are of any importance whatever in vast tracts of modern mathematics."[7] As Bell saw it, their importance lies in computation, which comes only after the mathematics is done, and where the decimal numerals have been serving well—certainly better than the Roman numerals had done.

If elementary and secondary teachers were being retrained to teach nondecimal numeration, it also made sense to modify the original training of teachers. The Committee on the Undergraduate Program in Mathematics recommended in the early 1960s "that decimal and nondecimal systems be studied" by future elementary teachers.[8]

These various recommendations were not ignored—at least not by the publishers—as will be seen in the survey of selected textbooks included in Chapter 8. Perhaps the recommendations were mere symptoms of the revolution and not causes of it, as must be reckoned the continued publicity enjoyed by the binary system as a key to computers and other information machines. The mid-1960s seemed to bring a crescendo of such public references, two of which shall be mentioned in particular here. One was *Time* magazine's account of the Mariner IV mission that resulted in the first photographs of Mars being transmitted to Earth.

> Each picture was made up of 200 lines—compared with 525 lines on commercial TV screens. And each line was made up of 200 dots. The pictures were held on the tube for 25 seconds while they were scanned by an electron beam that responded to the light intensity of each dot. This was translated into numerical code with shadings running from zero for white to 63 for deepest black. The dot numbers were recorded in binary code of ones and zeros, the language of computers. Thus white (0) was 000000, black (63) showed up as 111111. Each pic-

ture—actually 40,000 tiny dots encoded in 240,000 bits of binary code—was stored on magnetic tape for transmission to Earth after Mariner had passed Mars. More complex in some respects than the direct transmission of video data that brought pictures back from the moon, the computer code was necessary to get information accurately all the way from Mars to Earth.[9]

The other was the Bell System exhibit at the New York World's Fair of 1964–65, that featured a number translator about the size of a pinball machine. One could push buttons for, say 1964, and the machine would display not only '1964' but also its binary equivalent, 11110101100, and its equivalent in Roman numerals, MCMLXIV. A sign on this machine declared:

> The number translator is an example of what takes place each time you dial a call. Telephone equipment translates the number dialed to simple "1"'s and "O"'s which operate other switches to convert your call.[10]

Although this information is somewhat misleading, as will be shown in Chapter 7, the idea that strings of binary digits are involved is correct. Bell repeated this same message on a nearby wall.

yes
no
1
0
on
off
the electronic
language of
computers and
communications
machines

The 1960s also saw many parents attending evening school to learn some of the new school mathematics, which usually included a heavy portion of nondecimal numeration. Also the parents could choose from a number of books, such as E. Begle's *Very Short Course in Mathematics for Parents*. This book devotes 21 of its 53 pages to nondecimal numeration, but is unusual in not including the words "new" or "modern" in its title.

A careful survey of existing publications indicates that no historically oriented, self-contained overview of modern nondecimal numeration systems exists. Such an overview is needed, not to give the topic further momentum, but to lay the foundation of sound judgment about the extent and quality of its penetration into school mathematics. As Friedrich Unger wrote:

> He who wants to become master in his field should study its history. Without historic foundation all knowledge remains incomplete and the judgment about appearances of the present unsure and immature.[11]

This work is a library study concerned with the literature on numeration systems, primarily nondecimal ones. It deals with *standard* numeration systems, i.e., those patterned exactly after our common Hindu-Arabic system with ten's role taken over by some base β, where β is any whole number greater than one. Only those nonstandard systems that have found applications in computers or other information machines—or that are otherwise particularly relevant—are included. The study covers the period from the latter part of the sixteenth century to the present. Some of the earlier literature on this topic is in foreign languages and calls for resumés in English. Many of the publications prior to 1870 are difficult to obtain.

A *numeration system* is a system of number representation. It shall be called *standard* if it is our common Hindu-Arabic system or one exactly patterned after it, except for giving the special role of ten to another positive integer greater than one.

A *standard numeration system* is thus characterized by:

1. having β basic symbols, called *digits*, one for each of the β integers 0, 1, ... , (β-1).
2. having a whole number N greater than (β-1) represented

by a *string* of digits, $a_n a_{n-1} \ldots a_1 a_0$, this string being an abbreviation for

$$a_n\beta^n + a_{n-1}\beta^{n-1} + \ldots + a_1\beta^0$$

which is known as the *expanded form* (to the base β) of the number N.

Our common Hindu-Arabic system, which has $\beta = 10$, shall usually be referred to as the *decimal* or *base 10* system. In the string of digits, $a_n \ldots a_0$, each digit a_i has the *place value* β^i associated with it. The number β is known as the *base* or *radix* of the system. If the context fails to make clear which base is intended, β can be identified (in words or decimally) as a subscript at the end of the string, thusly,

$$N = (a_n \ldots a_0)_\beta$$

It is convenient but not essential to use the first β Hindu-Arabic numerals for digits, these can then be augmented by letters for β greater than ten. When each digit is itself a compound symbol, commas are inserted between digits, resulting in the usual notation from the field of number theory:

$$N = (a_n,\ a_{n-1},\ \ldots\ ,\ a_1,\ a_0)_\beta$$

it is being understood that each digit is represented decimally.

To illustrate the notation,

$$98 = (1\ \ 100\ \ 010)_2 = (1{,}38)_{60}$$

and if K = the number of seconds in one day, then

$$k = (24{,}0{,}0)_{60} \text{ and } k\text{-}1 = (23{,}59{,}59)_{60}$$

Among the principle sources of material for this study were Herbert McLeod's, William Schaaf's, and Lewis Seelbach's bibliographies.[12] McLeod's was restricted to articles that had appeared in scholarly journals, in whatever language, during 1771–1900. It had 47 items under "Scales of Notation." Schaaf's had twice that many, and included many books as well as articles. All of Schaaf's material is in English, and most of it was published after 1900, from sources like *The*

Mathematics Teacher and *School Science and Mathematics*. Seelbach's "Duodecimal Bibliography" was the most catholic in taste and the only one to be annotated. It included publications of many types and in many languages from 1585–1952. Fortunately the Seelbach bibliography, despite its title, did not restrict itself entirely to the duodecimal system. Of the over 350 publications listed therein, some were there by virtue of containing a single line favorable or unfavorable to duodecimals, while others contained over 50 pages of serious discussion of numeration systems in general.

Many of the items in the above three bibliographies provided additional references. Especially fruitful were the books by Wilhelm Ahrens and Leonard Dickson.[13] Ahrens's chapter on numeration systems represents the most serious previous attempt at a self-contained, historically oriented survey of modern numeration systems. Dickson's first volume of *History of the Theory of Numbers* does not treat numeration systems as a separate topic, but nevertheless provides many references—especially in Chapter VI ("Periodic Decimal Fractions; Period Fractions; Factors of $10^n \pm 1$") and Chapter XX ("Properties of the Digits of Numbers").

Two smaller works, one unpublished, give a wealth of references out of proportion to their size. One is Raymond Archibald's article in the *American Mathematical Monthly* of March, 1918, the other a spirit-duplicated study guide which Jones had prepared.[14]

The sources used for materials involving applications to computers will be identified in Chapter 7.

NOTES TO CHAPTER I

[1] W. W. Sawyer, *Visions in Elementary Mathematics*, (Baltimore: Penguin Books, 1964), pp. 25–26.

[2] Phillip S. Jones, "Historically Speaking, —" *The Mathematics Teacher* 46 (December 1953):575.

[3] Commission on Mathematics, *The Mathematics of the Seventh and Eighth Grades*, (New York: College Entrance Examinations Board, 1957), p. 2.

[4] School Mathematics Study Group, *Experimental Units for Grades Seven and Eight*, (New Haven: Yale University, 1959), T–II–1.

[5] Daniel D. McCracken, *A Guide to FORTRAN Programming*, (New York: Wiley, 1961), p. 4.

[6] Daniel D. McCracken, *A Guide to FORTRAN IV Programming*, (New York: Wiley, 1965), p. 30.

[7] E. T. Bell, *The Development of Mathematics*, (New York: McGraw-Hill, 1945), p. 52.

[8] Committee on the Undergraduate Program in Mathematics, *Course Guides for the Training of Teachers of Elementary School Mathematics*, third draft, (Berkeley: Committee on the Undergraduate Program in Mathematics, August 1963), p. N6.

[9] "Space Exploration," *Time* 86 (July 23, 1965):43.

[10] This and other information concerning the Bell System exhibit is based upon personal observation.

[11] Friedrich Unger, *Die Methodik der Praktischen Arithmetik*, (Leipzig: B. G. Teubner, 1888), p. ii.

[12] Herbert McLeod, ed., *The Subject Index of the Royal Society of London Catalogue of Scientific Papers*, 3 vols. (Cambridge: Cambridge University Press, 1908–14), vol. 1: *Mathematics*, pp. 45–46; William L. Schaaf, "Scales of Notation," *The Mathematics Teacher* 47 (October 1954):415–417; Lewis Carl Seelbach, "Duodecimal Bibliography," *Duodecimal Bulletin* 8 (October 1952):1–65.

[13] Wilhelm Ahrens, *Mathematische Unterhaltungen und Spiele*, 2 vols. (Leipzig: B. G. Teubner, 1910); Leonard Dickson, *History of the Theory of Numbers*, 3 vols. (New York: G. E. Steckert and Company, 1934).

[14] Raymond Clark Archibald, "The Binary Scale of Notation, a Russian Peasant Method of Multiplication, the Game of Nim and Cardan's Rings," *The American Mathematical Monthly* 25 (March 1918):139–142; Phillip S. Jones, "Numeration or Scales of Notation," (Unpublished study guide for a course at the University of Michigan, Ann Arbor, c. 1964).

·II·

BEFORE LEIBNIZ

THOMAS HARIOT[1] (1560–1621)

The English mathematician and astronomer, Thomas Hariot, left several thousand pages of unpublished manuscripts.[2] On one of these there appears, without comment, the following:

		16	16
1	1	17	16+1
2	2	18	16+2
3	2+1	19	16+2+1
4	4	20	16+4
5	4+1	21	16+4+1
6	4+2	22	16+4+2
7	4+2+1	23	16+4+2+1
8	8	24	16+8
9	8+1	25	16+8+1
10	8+2	26	16+8+2
11	8+2+1	27	16+8+2+1
12	8+4	28	16+8+4
13	8+4+1	29	16+8+4+1
14	8+4+2	30	16+8+4+2
15	8+4+2+1	31	16+8+4+2+1

The next hundred pages of the manuscript show a preoccupation with various techniques of tabulating all combinations (nonempty subsets) of n things—for n = 1, 2, 3, 4, and 5—followed by a prominent "etc." At first such tabulations were organized as follows:

a	1

a	
b	3
ab	

a	
b	
c	
ab	7
ac	
bc	
abc	

a	
b	
c	
d	
ab	
ac	
ad	
bc	15
bd	
cd	
abc	
abd	
acd	
bcd	
abcd	

a	
b	
c	
d	
e	
ab	
ac	
ad	
ae	
bc	
bd	
be	
cd	
ce	
de	
abc	31
abd	
abe	
acd	
ace	
ade	
bcd	
bce	
bde	
cde	
abcd	
abce	
abde	
acde	
bcde	
abcde	

Hariot's second tabulating technique came a few pages later. Again, he displayed all cases from n = 1 to n = 5, followed by "etc." To illustrate his new technique, it suffices to show the case of n = 3:

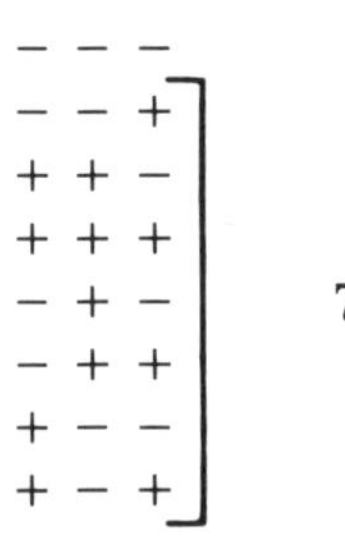

Apparently Hariot imagined the columns headed by a, b, and c, interpreted "+" as "yes, this letter is included in the combination," and "−" as "no, it is not." While Hariot's new technique brought a tabulation of 8 subsets of a set having 3 members, he preferred to count only the 7 nonempty ones or at least his bracket and his "7" seem so to indicate.

Eventually Hariot gravitated toward a third tabulating technique:

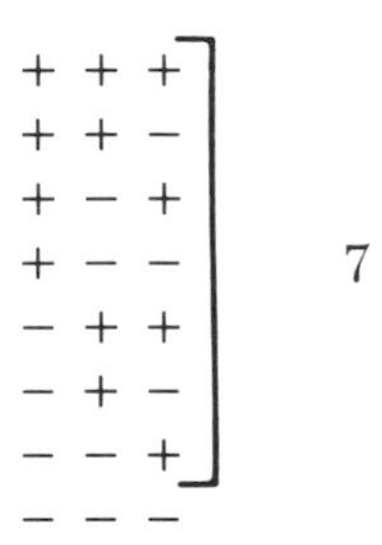

Again, the "7" and the bracketing are Hariot's. If we now interpret as above and furthermore interpret a = 4, b = 2, and c = 1 and assign to each of the 8 subsets that number which is the sum of its elements, then we find the subsets listed in good order from 7 to 0.

Hariot's preoccupation with the numbers 1, 3, 7, 15, and 31 (each of the type 2^n-1) as being associated with sets having 1, 2, 3, 4, and 5 members respectively, and his vigorous "etc.'s," suggest that he was aware of

Theorem 2.1: *There are* 2^n-1 *combinations (nonempty subsets) of* n *things.*

As he had also displayed how each of the first 31 natural numbers can be expressed as the sum of some combination of the 5 powers of 2 (1, 2, 4, 8, and 16), he apparently knew

Theorem 2.2: *The natural numbers from* 1 *to* 2^n-1 *can be expressed as the sum of some combination of the first* n *members of the set* {1, 2, 4, 8, 16, ...}.

Whether he followed this path or some other, the fact is, other pages of his manuscript show that he knew what we today call the binary numeration system or "base two."[3] He displayed, for example, 1101101, as the binary equivalent of 109, since binary $1101101 = 1 \cdot 64 + 1 \cdot 32 + 0 \cdot 16 + 1 \cdot 8 + 1 \cdot 4 + 0 \cdot 2 + 1 \cdot 1 = 64 + 32 + 0 + 8 + 4 + 0 + 1 = 109$. He gave examples of addition, multiplication, and subtraction with every number expressed in binary notation.

Before 1951, when J. W. Shirley reported on his examination of unpublished Hariot manuscripts (of ca. 1600), it was either Leibniz (1703) or Bishop Juan Caramuel y Lobkowitz (1670) who was given credit for having first discovered the binary system. English language references, however, have generally seemed unaware of Lobkowitz's claim.

Simon Stevin (1548–1620)

Edouard Lucas wrote in 1891:

> Simon Stevin of Bruges (died in 1633) had at one time proposed the duodecimal system of numeration, to match more nearly with our way of counting the months of the year, the hours of the day, and the degrees of the circumference; but the change of the system would actually produce too much inconvenience in relation to the small advantages which would result from *twelve* as the base.[4]

Ahrens mentioned Stevin as a duodecimal advocate in 1901.[5] However, my examination of the applicable portions of Stevin's *Les oeuvres mathématique* yielded no confirmation. Ahrens eventually concluded that Lucas had been in error and indicated so in the 1910 edition of his book.[6]

Even if it should be proven that Lucas's claim has some basis in fact, there is no doubt that Stevin's dominant efforts went toward further *decimalization*. He extended decimal numeration to fractions and even decimalized the system of weights and measures. Stevin certainly succeeded in the former. D. J. Struik reports that "Decimal fractions became a regular part of the curriculum in arithmetics as a result of the 1585 *De Thiende* by Simon Stevin."[7]

It was not for lack of effort that there was to be a 200-year delay in reaching the latter aim in continental Europe, for Stevin sternly demanded that the governments of Europe decimalize immediately, otherwise "a future generation would surely not pass up so great an advantage."[8]

Francis Bacon (1561–1626)

According to David Kahn's *The Codebreakers*,[9] Bacon published his *De Augmentis Scientarum* in 1623. In this work appears what Bacon called a "bi-literal" code for 24 letters of the alphabet (i and v were also serving in the place of j and u at that time), but what today would be called a 5-bit code. The letter t, for example, was assigned to the 5-bit string 10010, or rather BAABA, since Bacon used A and B instead of 0 and 1. Similarly, a = AAAAA = 00000. Moreover, Bacon used precisely the strings from 00000 (0) to 10111 (23) in their proper numerical order. Further discussion of this appears in Chapter 7.

Claude-Gaspar Bachet (1581–1638)

According to Underwood Dudley, Bachet's 1624 (2nd) edition of *Problèmes plaisants et délectables* first introduced "Bachet's Problem of the Weights," namely:

> Find a series of weights with which one can make all weighings in integer numbers from one to as far as the sum of the weights.[10]

Many later writers, Leibniz (1703) and Barlow (1811) among them, connected the solution to this problem with bases 2 and 3. More of this later.

Blaise Pascal (1623–1662)

"De numeris multiplicibus," occupying 13 pages in Pascal's *Oeuvres*,[11] was presented to the Academie Parisienne in 1654 and first published in 1665 as a supplement to *Traité du triangle Arithmétique*. Johannes Tropfke, who was of course unaware of Hariot's work, called this 13-page paper "the first scientific treatment of other number systems than the decimal."[12]

As its title promises, the paper deals with divisibility properties of numbers deduced from the sum of their digits. Pascal began by stating that "nothing is better known in arithmetic than the proposition according to which any multiple of 9 is composed of digits whose sum is also a multiple of 9." He then proceeded to give several examples, after which he continued:

> As much as this rule is commonly used, I do not believe that anyone up to the present has given a demonstration of, or has even searched for, a generalization of this principle. In this paper, I will justify the divisibility rule for 9 and several similar rules; I will also reveal a general method which permits one to know by simple inspection of the sum of its digits, if a given number is divisible by another number, whatever it be; this method applies not only to our decimal system of numeration (a system established not out of natural necessity, as is commonly thought, but by convention, a rather poor one at that) but to any system of numeration of whatever base.

With the preliminary remarks thus ended, Pascal stated, in substance:

Theorem 2.3: *The number* $N = a_n \ldots a_0$ *is divisible by* K *if and only if the test number* T *is divisible by* K, *where*

$$T = a_0 + a_1 R_1 + \ldots + a_n R_n$$

and where the R_i's *are found as follows*:

Divide K *into* 10	*to obtain the remainder* R_1
Divide K *into* $10R_1$	*to obtain the remainder* R_2
...	
Divide K *into* $10R_{n-1}$	*to obtain the remainder* R_n

In the awkward notation of his day, Pascal required two pages for stating this theorem before proceeding to a proof of the same length. He first noted that if $N = a_0$, then $T = a_0$ and the theorem is self-evident. After covering the cases for 2 and 3-digit numbers, he remarked that "The demonstration would be the same if the given number were composed of more than three digits." Below is the substance of Pascal's proof for $N = a_1 a_0$:

Proof of Theorem 2.3: (case $N = a_1a_0$): Given that N is a multiple of K, i.e., $N = a_0 + 10a_1 = Kp$ for some integer p. Since $10 = Kx + R_1$ and $R_1 = 10 - Kx$ for some integer x, it follows that $T = a_0 + R_1a_1 = a_0 + (10 - Kx)a_1 = a_0 + 10a_1 - Kxa_1 = Kp - Kxa_1 = K(p - xa_1)$, i.e., T is a multiple of K. Given on the other hand, that T is a multiple of K, i.e., $T = Kg$ for some integer q, it follows that $T = a_0 + a_1(10 - Kx) = N - Kxa_0 = Kq$ and hence $N = K(q + xa_0)$, i.e., N is a multiple of K.

In effect, Pascal recommends that T be thought of as

$$T = a_0R_0 + a_1R_1 + \dots + a_nR_n$$

with $R_0 = 1$. This leaves the substance of the theorem and its proof intact and has, as will be seen, more than the evident mnemonic advantage. Pascal points out that the R_i's depend only on K and not N. For example, for K = 7 the following R_i's result:

i	0	1	2	3	4	5	6	7	8	9	10	11	. . .
R_i	1	3	2	6	4	5	1	3	2	6	4	5	. . .

where 132645 keeps repeating.

Pascal then spends more than a page to illustrate the theorem for N = 287542178 and K = 7. He shows that T(N) = 119 and that a second application would yield T(119) = 14. Should one fail to recognize this as a multiple of Y one could apply the theorem a third time to give T(14) = 7, which is evidence of N being a multiple of 7. The next four pages of Pascal's paper give further examples, although all are still restricted to decimal numeration. Among the facts developed there are these:

	R_i for i = 0,1,2,3,...
For K = 6	1, 4, 4, 4, . . .
K = 3	1 1 1 1 . . .
K = 9	1 1 1 1 . . .
K = 4	1 2 0 0 0 0 0 . . .
K = 8	1 2 4 0 0 0 0 . . .
K = 16	1 10 4 8 0 0 0 . . .

all of which result in expressions for T that are simpler than in the case of $K = 7$ and are especially simple for $K = 3$ and $K = 9$. Finally Pascal writes:

> The divisibility properties of numbers deduced from the sum of their digits rests simultaneously on the inherent nature of numbers and their representation in the decimal system of numeration. In all other systems, for example in the duodecimal system (a most convenient one indeed) which aside from the first nine digits,[13] uses two new symbols, in order to designate the number 10 with the one and 11 the other, in this mode of numeration, it would no longer be true that all numbers whose digit sum is a multiple of 9 is itself a multiple of 9.
>
> But the method that I have made known and the demonstration which I have given are as suitable to this system as to any other.

Pascal indicates that if one wished to know the R_i's in this duodecimal system, one would interpret "10" as *twelve* (no longer *ten*) and "30" as *three times twelve* and the following R_i's would result: $R_i = 1, 3, 0, 0, 0, \ldots$, i.e., one would have to examine only the last two digits of N and merely test $T = a_0 + 3a_1$ for divisibility by 9. Pascal concluded:

> One would also know that, in this same system of numeration, every number whose digit sum is a multiple of eleven is itself a multiple of eleven.
>
> In our decimal system in contrast, to test for divisibility by eleven, it would be necessary that the sum formed by the last digit, reduced by the next-to-last, then the preceding, reduced by the preceding, etc., be a multiple of eleven.
>
> It would be easy to justify these two rules and to obtain some others. But if I have touched on this subject at all, it is because I would gladly yield to the lure of novelty, but I restrain myself, lest I tire the reader by too much detail.

Table 1 contains Pascal's announced results stated as numbered theorems in analytic form for the convenience of later referral. The first one, Theorem 2.4, is the principal new result while the remaining ones are special cases of it. At least one of these (Theorem 2.8) admittedly predates Pascal.

TABLE 1

Divisibility Theorems Referred to by Pascal in his "De Numeris Multiplicibus" of 1665

Given that $N = (a_n \ldots a_o)_\beta$, $R_o = 1$, and that R_i is the remainder when K is divided into βR_{i-1} for each $i = 1, 2, \ldots, n$, then N is a multiple of K if and only if T is a multiple of K, where

Theorem No.	T	β	K
2.4	$a_0R_0 + a_1R_1 + \ldots + a_nR_n$	all	all
2.5	$a_0 + \beta^a a_1 + 2a_2 + 6a_3 + 4a_4 + 5a_5 + a_6 + 3a_7 + 2a_8 + 6a_9 + 4a_{10} + 5a_{11} + \ldots$	10	7
2.6	$a_0 + 4a_1 + 4a_2 + 4a_3 + \ldots$	10	6
2.7	$a_0 + a_1 + \ldots + a_n \equiv T_s$	10	3
2.8	T_s	10	9
2.9	$a_0 + 2a_1$	10	4
2.10	$a_0 + 2a_1 + 4a_2$	10	8
2.11	$a_0 + 10a_1 + 4a_2 + 8a_3$	10	16
2.12	$a_0 + 3a_1$	12	9
2.13	T_s	12	11
2.14	$a_0 - a_1 + a_2 - \ldots + (-1)^n a_n \equiv T_a$	10	11

NOTE: T_s and T_a are defined by the identities shown in the table and may be called *simple* and *alternating* digit sum, respectively.

Juan Caramuel y Lobkowitz (1606–1682)

The chapter entitled "Meditatio" in his *Mathesis biceps* of 1670 gives evidence that the learned Bishop Juan Caramuel y Lobkowitz did indeed yield to the "lure of novelty."[14] Two and a half pages are devoted to binary arithmetic. These are followed by separate treatments of bases 3, 4, 5, 6, 7, 8, 9, 10, 12, and 60. Base 2 had (like any possible base) been implicit in Pascal's paper, although certainly not displayed. Hariot's manuscripts were not yet published, while Bacon's work was "binary" only through suitable interpretation, so the honor of the first *publication* explicitly on binary arithmetic goes to Caramuel.

Tropfke considered "Meditatio" to be scientific and independent of Pascal's work.[15] Moritz Cantor judged *Mathesis biceps* to follow only well-worn paths except for the "Meditatio" chapter.[16] In contrast, Pascal's *Traité du triangle Arithmétique* overshadowed the appended paper on nondecimal numeration.

Caramuel's contemporaries took so little notice of his work that Leibniz was hailed as the discoverer of the binary system upon publication of his "Explication" 33 years later in 1703. Many investigators still cite this paper as the *first* published work on the topic.[17] This technical error is justified in substance, for the test of true publication is the existence of reaction and follow-up, a test failed by Caramuel's work.

At the beginning of "Meditatio" Caramuel asks:

> Is there one arithmetic or many? If many, what are they and how does one distinguish among them? Are they useful or only speculative? Or necessary? What place should they occupy in the order of things?

In article I on binary arithmetic, he tabulated the binary and decimal representations for the numbers from 0 to 32. Since he uses "a" for "1" in the binary representation (unlike Hariot before and Leibniz after him) a0a0 = 10 and a0aa = 11. In a context involving only binary and decimal numeration, his notation has the obvious advantage that one cannot mistake a binary string for a decimal one.

While contrasting the place values of the binary with those of the decimal system of numeration, Caramuel notes that the set of differences between successive members of {1, 2, 4, 8, ... } is again that very set, whereas the set {1, 10, 100, 1000, ... } generates {9, 90, 900, ... } as the corresponding difference set, i.e., it seems to him that doublings are more natural than ten-fold increases. In further support of this point he discusses the musical scale: do, re, mi, ..., do, re, mi, ..., where successive "dos" involve ratios 1:2:4:8: and so on.

He gives a table of powers of 2, the first and last entries being:

1	0
1 125 899 906 842 624	50

Caramuel describes the columns as containing "natural numbers" and their corresponding "logarithms (artificial numbers)."

The note at the end of the article ranges far afield and then concludes that while musicians usually are limited to a 3 to 5 cycle range (octaves), binary arithmetic cannot be limited to so small a number of cycles (doublings), but must go on without end.

In the next 11 articles, each devoted to another base, Caramuel never tires of giving a table of the powers of β (except for $\beta = 10$) and does so at least up to the 7th power. In the case of $\beta = 3$, having displayed up to the 18th power of 3, namely 387 420 489, he notes that this power may be obtained by squaring the 9th power, 19 683. He actually shows the full multiplication procedure (in decimal arithmetic) of 19683 × 19683, partial products and all. Yet this amounts to nothing more than the fact that $(3^9)^2 = 3^{18}$. He goes through a similar, long treatment with the table of powers of 7, showing in effect only that $(7^4)^2 = 7^8$.

In each article, he treats the reader to an elaborate account of what comes to his mind when he thinks of the number β. For $\beta = 3$ this includes the Holy Trinity, for $\beta = 4$ that the Pythagoreans seemed to favor 4, and for $\beta = 5$ that human beings have five fingers on one hand. The number 4 has the longest commentary with almost 3 pages, 5 the shortest with only half a page.

For $\beta = 4$, Caramuel shows 27 = aei, since he uses 0, a, e, and i as the four quarternary digits. He shows no nondecimal base β representations after $\beta = 4$ until $\beta = 60$, which is the only nondecimal base for which he shows addition problems, such as the following one:

$0'''$	$24''$	$36'$	52	$49'$	$56''$	A
	58	47	54	38	42	B
	62	83	106	87	98	C
1	3	24	47	28	38	D

Here he uses decimal representations for the sixty sexagesimal digits, whereas some later writers came to prefer a collection of symbols that included letters (capital as well as lower case), so that the digit symbols would be simple rather than compound. Row C shows some "over-loaded bases"; row D has Caramuel's final answer. Caramuel's errors have been preserved in the above example.

It is obvious that as a mathematician Caramuel was not Pascal. "Meditatio" is also disappointing in failing to show nondecimal multiplication, or even addition except for $\beta = 60$. Pascal's paper does not show this either, but compensates for it by giving a powerful general divisibility theorem good for all bases. Yet "Meditatio" has the honor of being the earliest publication actually to display nondecimal numeration for bases less than 10.

Erhard Weigel (1625–1699)

By the standards established by Hariot and Pascal, the mathematical content of Bishop Caramuel's treatment of base 4 is small, but Professor (of mathematics) Weigel's *Tetractyn* (1672), devoted exclusively to this base, offers little more.

What saves *Tetractyn* from a quick dismissal as being merely a proper subset of "Meditatio" is its special link with Leibniz, who studied under its author for half of the year 1663.[18] Leibniz, who was 17 years old at the time, was later to refer to this work in connection with his own work on nondecimal numeration. G. E. Guhrauer credited Weigel with having introduced the teenager to arithmetic, lower analysis, and combinatorics.[19] Florian Cajori saw Weigel's influence on some of the notation later chosen by Leibniz.[20] Cantor was impressed with Weigel's creative powers in elementary fields of mathematics, but nevertheless thought him unworthy of a professorship, if only for being ignorant of recent mathematical advances such as the work of Descartes. The fact that Weigel could attract more than 400 students to his lecture entitled "Astrognostich-heraldisches Collegium" (which had to be held outdoors for lack of a hall that could hold that many) did not dissuade Cantor from asking in disbelief, "and this man was Leibniz's teacher?"[21]

If *Tetractyn* outdoes "Meditatio" on base 4, it is not in mathematical content, but in the commentary. The following is a sample:

> Hierocles as well, after he had said that nature herself continually confines herself to the number four, and does not surpass that number in the greatest affairs, that is to say, in the elements, in the seasons, and other things of the year, . . .[22]

Since *Tetractyn* was published under the auspices of the newly formed *Societas Pythagorea*, Thomas Heath's comments are pertinent:

> To the Pythagoreans, 10 was the perfect number, since it was the sum of 1, 2, 3, and 4, the set of numbers called the "tetractys." This set of numbers includes the numbers out of which are formed the ratios corresponding to the musical intervals, namely 4:3 (the fourth), 3:2 (the fifth), and 2:1 (the octave). Such virtue was attached to the tetractys that it was for the Pythagoreans their "greatest oath" and was alternatively called "health." It also gives, when graphically represented by points in four lines one below the other,

.
. .
. . .
. . . .

a triangular number.[23]

Although Weigel repeatedly pointed with pride to *Tetractyn* as his major work, its mathematical content is practically limited to the display on page 11 shown here:[24]

1.	2.	3.	10.
11.	12.	13.	20.
21.	22.	23.	30.
31.	32.	33.	100.

Weigel urged adoption of base 4 numeration for common use and later proposed the German names ERFF, ZWERFF, and DREFF for quarternary 10, 20, and 30 respectively, and the names SECHT and SCHOCK for 100 and 1000.[25]

JOSHUA JORDAINE'S *Duodecimal Arithmetick* (1687)

Between two works of Weigel, published 21 years apart, there appeared *Duodecimal Arithmetick* by Joshua Jordaine. The author published the work himself in London and humbly arranged to have proceeds from its sale go directly to His Majesty's treasury. The 300-page book's preface begins:

> He that shall but duly consider, that Duodecimals are the Foundation of all our English Measures, will soon see that nothing can be more natural and genuine for the finding of Superficial and Solid Contents, than a Duodecimal Arithmetick . . . .

On his title page, Jordaine assures the wary reader that the rules of this arithmetic are made "Plain and Easie for the meanest Capacity."

To start with, Jordaine provides his readers with four tables to be used as references in duodecimal arithmetic. The first three give all multiplication facts (decimally) up to $12 \times 12 = 144$. Table No. 4 is as follows:

$$\left.\begin{array}{l}6\\4\\3\\2\\1\tfrac{1}{2}\\1\end{array}\right\} \text{ is } \left\{\begin{array}{l}1/2\\1/3\\1/4\\1/6\\1/8\\1/12\end{array}\right\} \text{ of } 12$$

Jordaine adds:

> Then having learnt the Three first Tables, and acquainted yourself well with the parts of the last, you must learn the use and practise of those parts; for which purpose you must diligently observe, and perfectly learn these following Rules, viz.
>
> Rule 1
>
> To multiply any given number consisting of Integers, Primes, Seconds, Thirds, etc. by any number of Primes that are an Aliquot part of 12.
>
> 1. Example
>
> 6 Primes is the 1/2. Multiply 14--08′--10″ By 6 Primes
>
> 7--04′--05″ Facit

The very first example shows that Jordaine fell short of *standard* duodecimal numeration, if only because of the "14" which exceeds the base. Had Caramuel earlier bothered to actually exhibit instead of merely discuss duodecimals, one suspects that *he* would have shown the number

14--08′--10″

as follows: 1′ 2 8′ 10″, consistent with what in fact he had done in the case of base 60.

Still earlier, if Pascal had gone beyond merely discussing base 12 representations, and if the symbols A and B had been chosen to mean ten and eleven respectively, he would surely have shown this very number more compactly as 12.8A, where the period is used as a duodecimal point or fraction marker. In this notation, which is consistent with Pascal's suggestions, Jordaine's Example 1 becomes:

$$12.8\text{A} \times 0.6 = 7.45$$

where Jordaine's readers are encouraged to note that duodecimal 0.6 is the same as 1/2. Indeed Rule 1 and certain later ones can be summarized as encouraging the use of facts derived from Table 4, such as:

$$6 \times 12^n = \frac{1}{2} \times 12^{n+1}$$

Jordaine's Rule 2 urges the reader to think of numbers that are not aliquot parts (not in Table 4) as the sum of 2 or more numbers which are. Thus 7 is to be thought of as 6 + 1, as the following example from p. 9 illustrates:

Multiply	10--10′--07″			By 7 Primes
	5	05	03	06
	0	10	10	07
	6	04	02	01	Facit

With the help of Pascal's notation, this is simply:

$$\begin{aligned} A.A7 \times 0.7 &= A.A7 \times (0.6 + 0.1) \\ &= 5.536 + 0.AA7 \\ &= \underline{\underline{6.421}} \end{aligned}$$

Few of Jordaine's numerous examples avoid overloading the unit's digit. Three of them, translated into Pascal's notation, follow:

Example 2, p. 13:

```
  0.BBB
  0.11
 ------
    BBB
   BBB
 ------
 0.10BAB
```

Example 1, p. 25: 8.89 / 4 = 2.223

Example 1, p. 13: 0.11 × 0.1 = 0.011

The overloading of the unit's digit reaches a high in Example 2 on p. 22: Multiply (6378--04--08″) By (349--06--09) to which, after many details, Jordaine supplied an answer of (2229645--06′--09″--06″)

Many of his rules are simple special devices to facilitate computation. For example, he urged that division problems of the type $N/(1/2)$ be handled as a multiplication problem, $2 \times N$. Similarly he asked his readers, in effect, to note that:

$$N/(1/3) = 3N$$
$$N/(5/8) = 8N/5$$

and that $$N/288 = (N/12)/6/4.$$

On page 50, Jordaine offered a procedure for converting a decimal fraction, say 0.862, into its equivalent duodecimal fraction. 0.862 is to be multiplied by 12 to yield 10.344, and from this product the integral part (10) is to be subtracted to yield 0.344. Now the process is to be repeated, i.e., 0.344 is to be multiplied by 12 to yield 4.128. Again this result is to be multiplied by 12 to yield 1.536, and so on. The integral parts 10, 4, 1, ... are the duodecimal digits, i.e.,

$$(0.862)_{10} = (0.A41\ldots)_{12}$$

This is Jordaine's only example of conversion from a decimal to a duodecimal fraction. He offers no justification for the procedure, but it may be noted that the problem is equivalent to finding the unknown duodecimal digits a, b, c, d, ... in the following equation:

(1) $0.862 = (0.abcd\ldots)_{12} = a/12 + b/(12^2) + \ldots$

When (1) is multiplied by 12 it becomes:

(2) $10.344 = (a.bcd\ldots.)_{12} = a + b/12 + \ldots$

making it obvious that $a = 10 = (A)_{12}$. After subtracting $10 = a$ from equation (2) the result is:

(3) $0.344 = (0.bcd\ldots)_{12}$

Each subsequent multiplication by 12 will serve to identify another one of the digits b, c, d, etc.

An analogous procedure for converting duodecimal to decimal fractions calls for repeated multiplication of the given fraction by *ten* and doing so "duodecimally." Jordaine gave only one example. Translated into Pascal's notation, it appears as follows. If the given duodecimal fraction is 0.843:

$$\begin{aligned} 0.843 \times A &= 6.A66 \\ 0.A66 \times A &= 9.750 \\ 0.750 \times A &= 6.220 \\ 0.220 \times A &= 1.980 \end{aligned}$$

Now the integral parts of the products serve to show that

$$(0.843)_{12} = (0.6961\ldots)_{10}$$

The foregoing is a resume of the first 50 pages of Jordaine's book. He goes on to cover mean proportion, square roots, cube roots, and the rule of three in the next 30 pages. Thereafter, i.e., pages 83–300, he shows the applications of duodecimal arithmetic to mensuration, "gauging" (which includes figuring how much lumber is needed to accomplish a certain task), and "cask gauging" (which includes finding the volume of conical and spheroidal casks).

None of the earlier writers that have been discussed dealt with *fractions* written to nondecimal bases, except possibly Caramuel, whose sexagesimal (24″ 36′ 52 49′ 56″) can be so interpreted. Did Jordaine realize that his duodecimal arithmetic could be extended to integers? If so, did he refrain from doing so because it might have interfered with its application to English measures? Such measures are not fully duodecimalized, since there is, for example, no unit length equal to a dozen feet.

WEIGEL'S *Philosophia Mathematica* (1693)

In this book, published in 1693 in Jena, Weigel repeated much of his earlier work on base 4.[26] A modest advance beyond his earlier work can be noted in larger tables, the outline form of the commentary, and the inclusion of some base 4 multiplication and division examples, such as the following:

232	"in brief"	232
23		23)13322
12	232	112
21	23	212
12	2022	201
10	1130	112
12	13322	112
10		000
13322		

This modest advance over his earlier work came too late to influence Weigel's now famous former student. At the age of 47 Leibniz had already done much more serious work with nondecimal numeration in his unpublished manuscripts of the preceding 15 years.[27]

NOTES TO CHAPTER II

[1]Hariot's name is variously spelled with rr and/or tt. For further information on Hariot see Shirley's *Thomas Harriot* (sic).

[2]Thomas Hariot, "Mathematical Calculations and Annotations," (Film 128 in the library of the University of Michigan, original manuscript in the British Museum), 2:29.

[3]J. W. Shirley, "Binary Numeration Before Leibniz," *American Journal of Physics* 19 (November 1951):452–54.

[4]Edouard Lucas, *Récréations Mathématique*, 4 vols. (Paris: Gauthier-Villars et Fils, 1891), 1:147.

[5]Wilhelm Ahrens, *Mathematische Unterhaltungen und Spiele*, (Leipzig: B. G. Teubner, 1901), p. 25.

[6]Wilhelm Ahrens, *Mathematische Unterhaltungen und Spiele*, (Leipzig: B. G. Teubner, 1910), p. 32.

[7]D. J. Struik, "Simon Stevin and the Decimal Fractions," *The Mathematics Teacher* 52 (October 1959):474.

[8]Simon Stevin, *Les oeuvres mathématique*, (Leyden: B. and A. Elsevier, 1618), 2:616. Nevertheless, Stevin is still often found prominently listed as a duodecimal advocate. Lewis Carl Seelbach, of the Duodecimal Society of America, so lists him in his "Duodecimal Bibliography," *Duodecimal Bulletin* 8 (October 1952):20, 57, citing as his only basis the same 1901 passage which Ahrens retracted in 1920.

[9]David Kahn, *The Codebreakers*, (New York: Macmillan, 1967), pp. 882–83.

[10]Underwood Dudley, "The First Recreational Mathematics Book," *Journal of Recreational Mathematics* 3 (July 1970):164–169.

[11]Blaise Pascal, *Oeuvres*, 14 vols. (Paris: Brunschvicg et Boutroux, 1908), 3:311–39. The paper occupies more than double the space if one counts the editor's introduction and the French translation which appears on odd-numbered pages. The full Latin title of the paper is "De numeris multiplicibus ex sola characterum numericorum additione agnoscendis," and the full French title is "De caractères de divisibilité des nombres deduit de la somme de leurs chiffres."

[12]Johannes Tropfke, *Geschichte der Elementar-Mathematik*, (Berlin: Walter de Gruyter und Co., 1937), 1:4.

[13]Pascal, as was customary in his day, tacitly assumes the modifer "non-zero," or he would, of course, have referred to the first *ten* digits.

[14]Juan Caramuel, *Mathesis biceps*, (Campaniae: L. Annison, 1670). The author's full name, Juan Caramuel y Lobkowitz, can be found either under C or L, the first name appearing at times as John or Johannis.

[15]Tropfke, *Geschichte der Elementar-Mathematik*, 1:4.

[16]Moritz Cantor, *Vorlesungen über Geschichte der Mathematik*, 4 vols. (Leipzig: B. G. Teubner, 1922), 2:771.

[17]Tropfke and Cantor are exceptions.

[18]Cantor, *Vorlesungen über Geschichte der Mathematik*, 3:29.

[19]G. E. Guhrauer, *Gottfried Wilhelm Freiherr von Leibniz*, 2 vols. (Breslau: Ferdinand Hirt, 1846), 1:26.

[20]Florian Cajori, "Leibniz, the Master-builder of Mathematical Notations," *Isis* 7 (1925):429.

[21]Cantor, *Vorlesungen über Geschichte der Mathematik*, 3:38.

[22]Erhard Weigel, *Tetractyn*, (Jena: Johann Meyer, 1672), p. 11.

[23]Thomas L. Heither, *A Manual of Greek Mathematics*, (Oxford: Clarendon Press, 1931), p. 42.

[24]Cantor, *Vorlesungen über Geschichte der Mathematik*, 3:39.

[25]Ibid., 3:40.

[26]Erhard Weigel, *Philosophia mathematica*, (Jena: Matthew Birckner, 1693), pp. 144–185.

[27]Although the Leibniz manuscripts are still unpublished, Dietrich Mahnke's 1912 report of his examination of them is available and will be reviewed later.

·III·

LEIBNIZ

LETTER TO THE DUKE (1697)

After wishing a Happy New Year to Rudolph August, Duke of Brunswick, Gottfried Wilhelm Leibniz (1646–1716) continued in his letter of January 2, 1697 as follows:

> And so that I won't come entirely empty-handed this time, I enclose a design of that which I had the pleasure of discussing with you recently. It is in the form of a memorial coin or medallion; and though the design is mediocre and can be improved in accordance with your judgment, the thing is such, that it would be worth showing in silver now and unto future generations, if it were struck at your Highness's command. Because one of the main points of the Christian Faith, and among those points that have penetrated least into the minds of the worldly-wise and that are difficult to make with the heathen is the creation of all things out of nothing through God's omnipotence, it might be said that nothing is a better analogy to, or even demonstration of such creation than the origin of numbers as here represented, using only unity and zero or nothing. And it would be difficult to find a better illustration of this secret in nature or philosophy; hence I have set on the medallion design IMAGO CREATONIS [in the image of creation].
>
> It is no less remarkable that there appears therefrom, not only that God made everything from nothing, but also that everything that He made was good; as we can see here, with out own eyes, in this image of creation. Because instead of there appearing no particular order or pattern, as in the common representation of numbers, there appears here in contrast a wonderful order and harmony which cannot be improved upon. Inasmuch as the rule of alternation provides for continuation, so that one can write without computation or the aid of

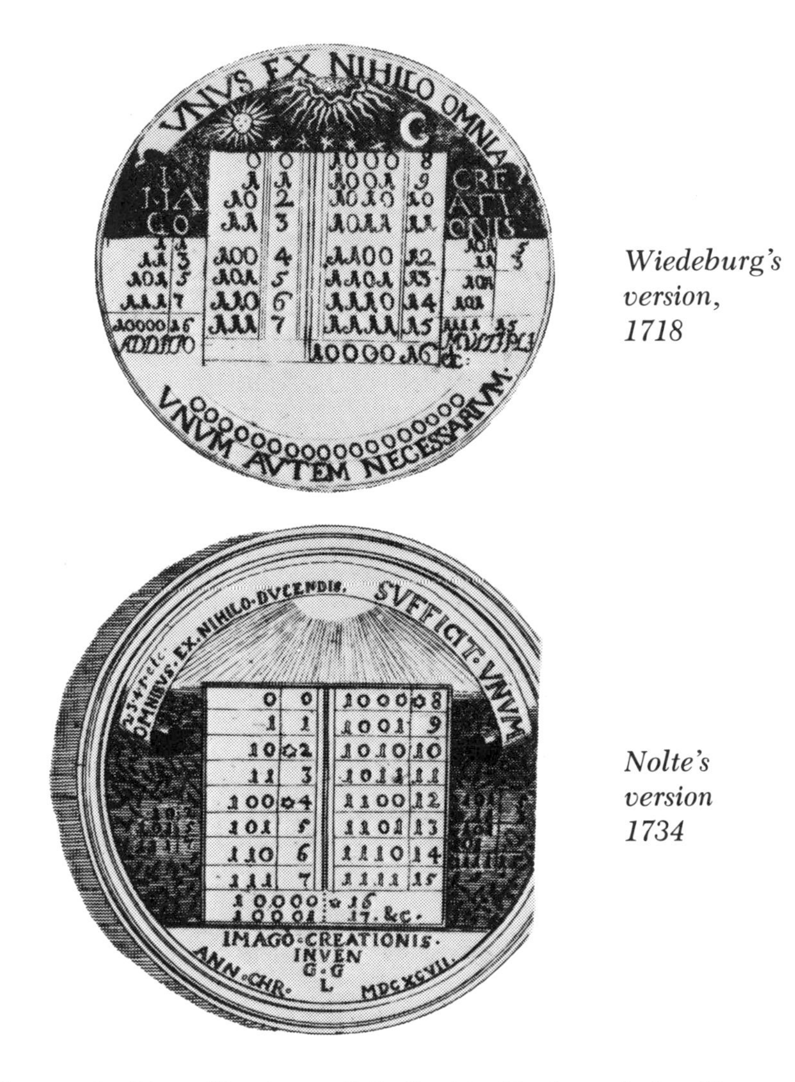

FIGURE 1. Two Versions of Leibniz's Design of the Binary Medallion. They are facsimiles of the ones appearing on the respective title pages of Johann Bernard Wiedeburg's *Dissertatio mathematica de praestantia arithmeticae binaria prae decimali* (Jena: Krebs, 1718) and Rudolf August Nolte's *Leibniz Mathematischer Beweis der Erschaffung und Ordnung der Welt in einem Medallion an den Herrn Rudolf August* (Leipzig: J. C. Langenheim, 1734.

memory as far as one wishes, if one alternates the last place 0,1,0,1,0,1, etc., putting these under each other; and then putting under each other in the second place (from the right) 0,0,1,1,0,0,1,1, etc.; in the third 0,0,0,0; 1,1,1,1; 0,0,0,0; 1,1,1,1; etc; in the fourth 0,0,0,0,0,0,0,0; 1,1,1,1,1,1,1,1; 0,0,0,0,0,0,0,0; 1,1,1,1,1,1,1,1; and so forth, the period or cycle of change becomes again as large for each new place. Such harmonious order and beauty can be seen in the small table on the medallion up to 16 or 17; since for a larger table, say to 32, there is not enough room. One can further see that the disorder, which one imagines in the work of God, is but apparent; that if one looks at the matter with the proper perspective, there appears symmetry, which encourages one more and more to love and praise the wisdom, goodness, and beauty of the highest good, from which all goodness and beauty has flown. I am corresponding with Jesuit Father Grimaldi, who is currently in China and also president of the Mathematics Tribunal there, with whom I became acquainted in Rome, and who wrote me during his return trip to China from Goa. I have found it appropriate to communicate to him these number representations in the hope, since he had told me himself that the monarch of this mighty empire was a lover of the art of arithmetic and that he had learned to figure the European way from Father Verbiest, Grimaldi's predecessor, that it might be this image of the secret of creation which might serve to show him more and more the excellence of the Christian faith.

So that I may explain the rest of the medallion I marked the main places, namely 10 or 2, 100 or 4, 1000 or 8, 10000 or 16, with * or an asterisk, because if one takes note of these, one is bound to see the origin of the remaining numbers. For example, why 1101 stands before 13 is shown by this demonstration:

1	1
00	0
100	4
1000	8
1101	13

and so it is with all other numbers. I have also given an example of addition and one of multiplication on the medallion at the sides of the table so that one could note from them the foundation of operations and how the arithmetic rules apply here

Gottfried Wilhelms Baron von Leibnitz
Mathematischer Beweis
Der
Erschaffung und Ordnung der Welt
In einem Medallion
An den Durchlauchtigsten Fürsten und Herrn,
Herrn Rudolph August,
Weyland
Regierenden Herzog zu Braunschw. und Lüneb. ꝛc.
entworfen,

und an das Licht gestellet
Von
RUD. AUG. NOLTENIO, Adv. Wolffenb.
Leipzig, zu finden bey Johann Christian Langenheim 1734

FIGURE 2. Facsimile of Title Page of Rudolf Nolte's 1734 booklet, in which Leibniz's 1697 letter to the Duke appears as well as the 1703 "Explication."

too. However, the intention is not to use this manner of reckoning in other than the study of and search for the secrets of numbers, and not at all for use in every-day life.

The next two paragraphs of the letter are devoted to a detailed description of the remaining features of the medallion design. Each of the two versions of Figure 1 is somewhat at variance with this description—Nolte's being the closer. The letter continues:

> A number written in this way will not be more than 4 times as long as in the conventional way. In it can be found, I think, so many wonderful and useful observations for the increase of scientific knowledge that the Mathematics Society of Hamburg [Hamburgische Rechnungsgesellschaft], whose industry and determination is praise-worthy, if some members thereof would wish to turn their thoughts and desires on it, would, I am certain, find such things, which would bring not inconsiderable renown to it [the Society] and to the German nation, because the matter was first brought to light in Germany. From this manner of writing numbers I see wonderful advantages accruing, which will subsequently also benefit common arithmetic.

Lest the idea of commemorating that wonderful binary system be insufficient motivation for having the medallion struck, Leibniz suggested that the other side be devoted in some manner to showing a bust of Rudolph August, the Duke of Brunswick.

This 1697 letter to the Duke was published by Heinrich Köhlern in 1720 in a booklet that also contained a Leibniz paper on Monadology. Köhlern seems to have appended the letter even though he had failed to see in it any "proof of the application of the binary system to the Creation."[1]

Rudolf Nolte had been much more impressed; he included the same letter (with minor deletions) in his 1734 booklet containing the "Explication" and also bearing a title page (Figure 2) that boldly announced Leibniz's mathematical *proof* of the Creation.[2]

The binary medallion apparently was never struck.[3] Nu-

0000000	0
0000011	3
0000110	6
0001001	9
0001100	12
0001111	15
0010010	18
0010101	21
0011000	24
0011011	27
0011110	30
0100001	33
0100100	36
0100111	39
0101010	42
0101101	45
0110000	48
0110011	51
0110110	54
0111001	57
0111100	60
0111111	63
1000010	66
1000101	69
1001000	72
1001011	75
1001110	78
1010001	81
1010100	84
1010111	87
1011010	90
1011101	93
1100000	96
1100011	99

etc.

FIGURE 3. Table from Leibniz's May 17, 1698 letter to Johann Christian Schulenburg.

merous writers have based a contrary assumption, in the last analysis, upon having seen some version of its *design*. The Duke was already 70 years old when he received the medallion proposal in 1697.

Two letters to Johann Christian Schulenburg (1698)

These letters were written by Leibniz on March 29 and May 17, 1698.[4] The first included merely an introduction to binary notation, but the second included the table of multiples of 3 shown in Figure 3

Focusing on the binary strings of this table, Leibniz made the following observations:

(1) The digit column at the right end of the binary strings is 01010101..., i.e., it involves the period 01.

(2) Each column (numbered from the right toward the left) has its own characteristic period:

1st column:	01
2nd column:	0110
3rd column:	00101101
4th column:	0001110011100011
5th column:	00000111110000011111100000111111
etc.	

(3) The period of the nth column is 2^n digits long.

(4) If the second half of each period is written under the first half, it becomes apparent that they differ only in having 1 replace 0 and vice versa:

3rd column:	0010
	1101
4th column:	00011100
	11100011
5th column:	0000001111100000
	1111110000011111

0	0	0	0	0	0	0
0	0	0	0	0	1	1
0	0	0	0	1	0	2
0	0	0	0	1	1	3
0	0	0	1	0	0	4
0	0	0	1	0	1	5
0	0	0	1	1	0	6
0	0	0	1	1	1	7
0	0	1	0	0	0	8
0	0	1	0	0	1	9
0	0	1	0	1	0	10
0	0	1	0	1	1	11
0	0	1	1	0	0	12
0	0	1	1	0	1	13
0	0	1	1	1	0	14
0	0	1	1	1	1	15
0	1	0	0	0	0	16
0	1	0	0	0	1	17
0	1	0	0	1	0	18
0	1	0	0	1	1	19
0	1	0	1	0	0	20
0	1	0	1	0	1	21
0	1	0	1	1	0	22
0	1	0	1	1	1	23
0	1	1	0	0	0	24
0	1	1	0	0	1	25
0	1	1	0	1	0	26
0	1	1	0	1	1	27
0	1	1	1	0	0	28
0	1	1	1	0	1	29
0	1	1	1	1	0	30
0	1	1	1	1	1	31
1	0	0	0	0	0	32

FIGURE 4. Table of Numbers from Leibniz's "Explication."

Correspondence With Johann (Jean) Bernoulli (1701)

An exchange of several letters in the period from April 5 to May 7, 1701 was devoted, in part, to the binary system.[5] Initially Leibniz offered an explanation not unlike that in his letter to the Duke. Indeed, nothing beyond the mathematical content of that letter to the Duke developed, except possibly Bernoulli's explicit recognition that the binary place values 1, 2, 4, 8 ... are in fact powers of 2 and that, for example,

$$1701 = 2^{10} + 2^9 + 2^7 + 2^5 + 2^2 + 1 = 11010100101.$$

"Explication" (1703)

"Explication de l'arithmétique binaire" by Leibniz appeared in the 1703 volume of the *Memoires de l'Academie Royale des Sciences* on pages 85–89. This explanation of binary arithmetic was the first publication on this topic to result in a significant impact on the scientific community.

Leibniz, now 57, had been a frequent contributor to the *Memoires* of this Parisian academy. The Berlin academy was not to begin its publications until 1710. The article begins:

> The ordinary reckoning of arithmetic is done by tens. One draws on ten characters, which are 0, 1, 2, 3, 4, 5, 6, 7, 8, 9, which signify zero, one and its successors up to nine inclusive. Upon coming to ten, one begins over again, expressing ten by 10, ten times ten or hundred by 100, ten times hundred or thousand by 1000, ten times thousand by 10000, and so on.
>
> But instead of this progression by tens, I have for many years used the most simple of all, which goes by twos, having found that it is conducive to the perfecting of the science of numbers. Thus I have used no other characters but 0 and 1, and upon coming to two, I begin again. This is why two is expressed here by 10, and two times two or four by 100, two times four or eight by 1000, two times eight or sixteen by 10000, and so on.
>
> Here is the Table [Figure 3.4] of Numbers in this fashion, which one could extend as far as one would wish. Here, one sees at a single glance the reason for a celebrated property of

Addition:

	Binary	Decimal
*	110	6
	111	7
	..	
	1101	13

Binary	Decimal
101	5
1011	11
....	
10000	16

Binary	Decimal
1110	14
10001	17
11111	31

Subtraction:

Binary	Decimal
1101	13
111	7
110	6

Binary	Decimal
10000	16
1011	11
101	5

Binary	Decimal
11111	31
10001	17
1110	14

Multiplication:

	Binary	Decimal
	11	3
**	11	3
	11	
	11	
	1001	9

Binary	Decimal
101	5
11	3
101	
101	
1111	15

Binary	Decimal
101	5
101	5
101	
1010	
11001	25

Division:

Decimal	Binary
15	~~1~~~~1~~~~1~~1
3	~~1~~~~1~~~~1~~1
	~~1~~1

Binary	Decimal
101	5

FIGURE 5. Examples of Arithmetic Operations Shown in Leibniz's "Explication."

the geometric progression by twos in whole numbers, that which permits the fact that if one has only these numbers of each degree, one can compose all the other whole numbers less than double the highest degree. [Compare Hariot, Theorem 2.2] Here, this is as if one said, for example, that 111 or 7 is the sum of four, of two, and one—and that 1101 or 13 is the sum of eight, four, and one. This property permits the Assayer to weigh all sorts of masses with few weights and could serve in coinage to give more value with fewer pieces.

These expressions of numbers being established would very much facilitate all sorts of operations. [Figure 3.5] All these operations are so easy that one would never need to try or guess, as is needed in ordinary arithmetic. One would no longer have to memorize as one must for ordinary reckoning, where it is necessary to know, for example, that 6 and 7 added together make 13 and that 5 multiplied by 3 gives 15 in accordance with the multiplication table. But here all this finds itself and proves itself at the start, as one sees in the examples preceded by signs * and **.

Nevertheless I do not at all recommend this manner of counting to replace the ordinary by ten. Aside from being accustomed to this and not needing to learn what one has already memorized, one finds the common counting by ten quicker and the numbers not as long. If one were to count by dozens or sixteens one would have even more of an advantage in this respect. But reckoning by twos, that is to say by 0 and 1, in recompense for its length, is more fundamental to science and gives new discoveries, which result in subsequent utility—even to the common way of numbering and above all for geometry. The reason is that the numbers, being reduced to their simplest principle, like 0 and 1, seem all around in the best possible order. For example, in the same Table of Numbers [Figure 3.4], one sees each column ruled by periods which always begin over again. In the first column this is 01, in the second 0011, in the third 00001111, in the fourth 0000000011111111, and so on. Little zeros have been put into the table to bring out better these periods. Also, lines have been put into the table, which confine such periods within them. It happens again that square numbers, cubes, and other powers, also triangular numbers, pyramid, and other figure numbers, have similar periods, of a sort that one can write the tables without calculation. One bit of tediousness in the beginning, which afterwards gives the means for economizing the calculation and going on to infinity by rule, is infinitely advantageous.

What is astounding in this reckoning is that this arithmetic by 0 and 1 happens to contain the secret of the lines of an ancient king and philosopher named Fohy, who is believed to have lived more than 4000 years ago, and whom the Chinese regard as the founder of their empire and their sciences. There are several linear figures which are attributed to him. They all come back to this arithmetic, but it suffices to show here the Figures of the Eight Cova, as they are called, which pass as fundamental, and to adjoin to them the explanation, which is manifest provided that one notices firstly that a whole line —— means unity or 1 and secondly, that a broken line -- means zero or 0.

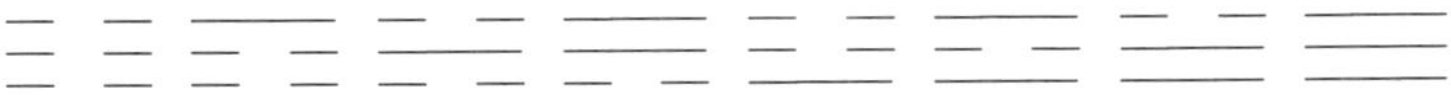

000	001	010	011	100	101	110	111
0	1	10	11	100	101	110	111
0	1	2	3	4	5	6	7

The Chinese lost the significance of these Cova or Lineations of Fohy, perhaps more than 1000 years ago. They have made commentaries on that, in which they have gathered I know not what far out meanings. They are of a sort that it is necessary that their true explanation now come to Europeans.

It is hardly more than two years ago that I sent to R. P. Bouvet, the celebrated French Jesuit, who died in Peking, my method of counting by 0 and 1. He needed nothing further to make the observation that this was the key to the Figures of Fohy. When thus he wrote me on November 14, 1701, he sent me this princely philosopher's Grand Figure, which goes to 64 lineations and leaves no room for doubt about the truth of our interpretation, which is such that one could say that this Father has deciphered the Enigma of Fohy with the aid of that which I had communicated to him. Since these Figures are perhaps the most ancient monument of science which exists on this earth, this restitution of their meaning, after so long an interval of time, would seem most curious.

The agreement between the Figures of Fohy and my Table of Numbers is seen better if the initial zeros are supplied, which may seem superfluous, but which serves better to mark the periods of the columns, as I have supplied them in effect with little circles, to distinguish them from the necessary zeros. This accord gives me a high opinion of the profundity of the meditations of Fohy, because that which seems easy to us now was not so in those far-removed times. The binary or dyadic arithmetic is, in effect, very easy today with little thought going into it, because our manner of counting is conducive to it; it seems that one cuts off only the excess of it. But this ordinary arithmetic by tens does not seem very ancient, at least the Greeks and the Romans had ignored it, and have been deprived of its advantages. It seems that Europe owes its introduction at the time of Pope Sylvester II, to Gerbert, who had seen it with the Moors of Spain.

Now, as one believes in China, that Fohy is also the author of ordinary Chinese characters, which were severely altered in subsequent times, his essay on arithmetic calls for this judgment: it might well be possible to uncover again some considerable things by way of the rapport between the numbers and the ideas, if one could unearth the foundations of this Chinese writing, which, much more than is believed in China, has consideration of numbers established in itself. R. P. Bouvet has strongly urged to push this point and to expect good results of this kind. However, I do not know if there was ever an advantage in this Chinese writing approaching that which ought to exist necessarily in the feature that I project. It is that all reasoning which one can pull from ideas might be pulled from their characters by a manner of reckoning, which would seem one of the most important aids to the human intellect.

Leibniz's "Explication" has hereby been given in full as translated from the French.

"Nouvelle Arithmetique" By Fontenelle (1703)

Bernard Le Bovier de Fontenelle published "Nouvelle Arithmetique" in the 1703 issue of *Histoire de l'Academie*

Royale des Sciences, pages 58–63. At the time, he was secretary of that Parisian academy. His unsigned article constituted an editorial comment on the "Explication" of Leibniz. They were contained in the same volume, for the *Memoires*, though separately paginated, were bound together with the *Histoire*.

Fontenelle pointed out that ten need not be the base of our arithmetic, and that indeed certain other bases would have advantages over it. Base 12, for example, would simplify dealings with certain fractions such as 1/3 and 1/4. He also noted that numbers have two sorts of properties, essential ones and those dependent on the manner of expressing them. As an example of the former he cited the property that the sum of the first n odd numbers equals n^2, and of the latter that a number divisible by 9 has a digit sum also divisible by 9. This same property would hold for 11 in the case of base 12. He reported that Leibniz had worked with the simplest of all possible bases, base two. This base was not recommended for common use because of the excessive length of its number representations, but Leibniz considered it particularly suitable for difficult research and as possessing advantages absent from other bases.

Fontenelle reported further that Leibniz had communicated this binary arithmetic in 1702, but had asked that no mention of it be made in the *Histoire* until he could supply an application. This application eventually came forth in the binary interpretation of the Figures of Fohy. The rest of Fontenelle's article is devoted to reporting that binary arithmetic was invented not only by Leibniz, but also by Professor Lagny at about the same time.

Thomas Fantet De Lagny (1660–1734)

Lagny was admitted to the Paris academy in 1695, contributed frequently to its *Memoires*, and published several books on scientific subjects. Numerous later references (some garbled) to Lagny's work on the binary system are, in the last analysis, traceable to the following passage from "Nouvelle Arithmetique" by Fontenelle:

If Mr. Leibniz did not discover binary arithmetic simultaneously with the Emperor Fohy, at least Mr. de Lagni discovered it simultaneously with Mr. Leibniz. Mr. Lagni, Professor of Hydrography at Rochefort, works, as one has already been able to see in the 1702 *Histoire*, at perfecting the science which he professes. For navigational use, he has developed a new trigonometry, and while investigating the entire system of logarithms, which had been invented principally for trigonometry, he had seen in it defects and inconveniences, which he had been able to remedy only by devising binary arithmetic. The great inconvenience of logarithms is the changing of multiplication and division, which are long and difficult operations for large numbers, into addition or subtraction, which are much simpler and easier. But Mr. de Lagni claims that this advantage, which the theory promises to fulfill so magnificently, is reduced to nothing in practice. On the contrary, these logarithms, which are sort of sham or artificial numbers, are only a detour on the way to arriving at *natural* numbers—the only ones being looked for. He calls to witness all those who have calculated by this method that additional work is involved, done more easily, perhaps, but requiring a longer time. He even maintains that these logarithms give false results for large numbers, and he cites as proof for this a computation that Henry Briggs had given as an example of the use of logarithms in his *Arithmetique Logarithmique* beginning on page 27.

In binary arithmetic, multiplications and divisions develop necessarily through simple additions and subtractions—no detour being necessary, such as the one through logarithms in common arithmetic—and accordingly all the advantage, which common arithmetic fails to get out of the use of logarithms, is inherent in binary arithmetic, whose multiplications and divisions Mr. Lagni for this reason calls *natural logarithms*. He has given his idea at length in his paper, which he presented this year at Rochefort and which he sent to the Academy. The little we have said about it, should suffice for the eyes of those who would wish to go deeper into this new arithmetic.

As even the greatest of mathematicians could very legitimately be desirous of the glory of a simultaneous discovery with Mr. Leibniz (without having been in his following), we feel obliged to testify here on behalf of Mr. Lagni, that, having

> always been at Rochefort, he would not seem to have had any knowledge of that which Mr. Leibniz had sent to the Academy on binary arithmetic.

Calling a number *natural* and its logarithm *artificial* was common at this time (Caramuel had done so, for example). Quite unusual, however, was the use of "natural logarithms" to mean "a *natural* aid to computation," i.e., "an aid to computation that *avoids* logarithms."

Correspondence With Jacques Bernoulli (1704–5)

Jacques or Jacob was the older brother of Jean, whose correspondence with Leibniz was mentioned earlier. During the last year of his life, Jacques exchanged several letters with Leibniz that were, in part, devoted to the binary system.[6] After Leibniz raised the topic of columnar periods in the binary strings of the integers 1, 2, 3 ..., Bernoulli showed the following table of squares:

0	0
1	1
100	4
1001	9
10000	16
11001	25
100100	36
110001	49
1000000	64
1010001	81
1100100	100
1111001	121
10010000	144
10101001	169
11000100	196
11100001	225
100000000	256
etc.	etc.

He pointed out that the first column has a period of 01, the second is all zeros, and the third has a period of 1000. To Bernoulli's comment that the fourth column had no obvious period, Leibniz supplied 10100000 for the fourth and 1101010110000000 for the fifth.

Bernoulli communicated the following string of binary digits

11 1100 1000 0001 0011 1111 0110 0110 0011 0110 1101 1001 1000 0100 1111 0010 0101 1001 0101 1011 0110 1010 0011 0111 0010 1010 0000 0011 1101 0000

as the supposed binary equivalent of

$$a = 3.14159\ 26535\ 89793\ 23846\ 26433\ 83279\ 50288,$$

where a is the 36-decimal digit approximation to the number π. Since

$$3\frac{1}{8} < a < 3\frac{1}{4}$$

which in binary notation reads

$$11.001 < a < 11.010$$

it becomes obvious that Bernoulli made an error more serious than leaving out the fraction marker. In fact, his 118-bit string seems to be the binary equivalent of the *integer* $(10^{35}a)$. Bernoulli also offered

42161 26122 23212 12122 24124 21211 22111 12121 21111 32132 11111 74114

as a description or abbreviation of that 118-bit string, where the "4" indicates 4 ones, the "2" indicates 2 zeros, then 1 one, 6 zeros, and so on.

Mahnke's Report on Leibniz Manuscripts

The following account is based entirely on a report by Dietrich Mahnke, who had examined unpublished Leibniz manuscripts.[7]

On September 12, 1680, Leibniz wrote the first known proof of

Theorem 3.1: *When* p *is a prime and* β *is not a multiple of* p, *then*

$$\beta^{p-1} = 1 \pmod{p}.$$

This is known as Pierre de Fermat's (1601–1665) theorem. Fermat is best known for his "Last Theorem," which has not been proven to this day. Leibniz had been dealing with binary and decimal fractions since 1677. He knew for example that 1/3 has the following binary representation

$$1/11 = 0.0101\overline{01}\ldots$$

with a 2-digit binary period and that 1/10 would have a 4-digit period, since the rising powers of 2 always leave the remainders 2, 4, 8, 6, in periodic order upon division by 10, and must repeat when all four of them have occurred. Obviously, at most n-1 different remainders can occur upon division by n, hence, a period in any base β can have at most n-1 digits. In unit fractions of the type 1/n, Leibniz viewed the dividend 1 as the first remainder and concluded that the second period begins when another remainder of 1 appears, at least when n is relatively prime to the base β.

At least one contemporary of Leibniz, John Wallis, knew that the lengths of the periods of the decimal equivalents of 1/n are at most n-1 and that they are frequently proper divisors of n-1. But Leibniz knew more precisely that the lengths are *always* divisors of n-1 when n is prime. He was also aware of a close connection with Fermat's theorem (3.1), a connection not to be rediscovered until Johann Heinrich Lambert did so in 1758.

The work of Leibniz seemed to be part of an effort to discover a prime number formula and a proof of the irrationality of the number π. He seemed to temporarily have succeeded in the latter when he argued that his formula

$$\pi/4 = 1 - 1/3 + 1/5 - 1/7 + \ldots$$

(known to him since 1674) indicated that π was irrational, since the decimal equivalents of the fractions would involve ever longer periods, hence, their sum, $\pi/4$, had to have an infinitely long period and be irrational. However, since the sum of infinitely many decimal fractions of ever longer periods may well be rational, the argument is invalid. For example:

$$1 + 1/3 + 1/9 + 1/27 + \ldots = 3/2$$

A valid proof of the irrationality of π, according to Mahnke, did not come until Lambert provided one in 1766.

Some Reactions to Leibniz's Binary Interpretation of the Figures of Fohy

The Figures of Fohy appear in the ancient Chinese book *I Ching* or *Book of Changes*. F. van der Blij has characterized this work as "a highly corrupt jumble of divination texts reminiscent of the oracle-bone documents of the second millennium B.C."[8]

James Norwood, on the other hand, described it as "a compilation of ancient wisdom that stretches back into prehistory, contains commentaries on sixty-four hexagrams the Figures of Fohy, any one of which its user may obtain by casting yarrow stalks (the traditional form) or tossing coins (the modern form)—moments of fate snatched from the ongoing river of change. The book is used not simply for fortune-telling but is, rather, oracular in the sense that it intends to convey wisdom."[9]

After Leibniz published his binary interpretation of the Figures of Fohy in the 1703 "Explication," no one was more delighted than Wilhelm Ernst Tentzeln, the editor of *Curieuse Bibliothec*, who published more details of this interpretation, based upon a private communication from Leibniz.[10] Tentzeln clearly considered it most curious that the supposedly intelligent Chinese had lost and then had failed to rediscover the meaning of these Figures, so that it took a European genius to do the job for them.

0	0	0	0	0	0	= 0
0	0	0	0	0	1	= 1
0	0	0	0	1	0	= 2
0	0	0	0	1	1	= 3
0	0	0	1	0	0	= 4
0	0	0	1	0	1	= 5
0	0	0	1	1	0	= 6
0	0	0	1	1	1	= 7
0	0	1	0	0	0	= 8
0	0	1	0	0	1	= 9
0	0	1	0	1	0	=10
0	0	1	0	1	1	=11
0	0	1	1	0	0	=12
0	0	1	1	0	1	=13
0	0	1	1	1	0	=14
0	0	1	1	1	1	=15
0	1	0	0	0	0	=16
0	1	0	0	0	1	=17
0	1	0	0	1	0	=18
0	1	0	0	1	1	=19
0	1	0	1	0	0	=20
0	1	0	1	0	1	=21
0	1	0	1	1	0	=22
0	1	0	1	1	1	=23
0	1	1	0	0	0	=24
0	1	1	0	0	1	=25
0	1	1	0	1	0	=26
0	1	1	0	1	1	=27
0	1	1	1	0	0	=28
0	1	1	1	0	1	=29
0	1	1	1	1	0	=30
0	1	1	1	1	1	=31
1	0	0	0	0	0	=32
1	0	0	0	0	1	=33
1	0	0	0	1	0	=34
1	0	0	0	1	1	=35
1	0	0	1	0	0	=36
1	0	0	1	0	1	=37
1	0	0	1	1	0	=38
1	0	0	1	1	1	=39
1	0	1	0	0	0	=40
1	0	1	0	0	1	=41
1	0	1	0	1	0	=42
1	0	1	0	1	1	=43
1	0	1	1	0	0	=44
1	0	1	1	0	1	=45
1	0	1	1	1	0	=46
1	0	1	1	1	1	=47
1	1	0	0	0	0	=48
1	1	0	0	0	1	=49
1	1	0	0	1	0	=50
1	1	0	0	1	1	=51
1	1	0	1	0	0	=52
1	1	0	1	0	1	=53
1	1	0	1	1	0	=54
1	1	0	1	1	1	=55
1	1	1	0	0	0	=56
1	1	1	0	0	1	=57
1	1	1	0	1	0	=58
1	1	1	0	1	1	=59
1	1	1	1	0	0	=60
1	1	1	1	0	1	=61
1	1	1	1	1	0	=62
1	1	1	1	1	1	=63

BINARY SYSTEM OF LEIBNITZ.

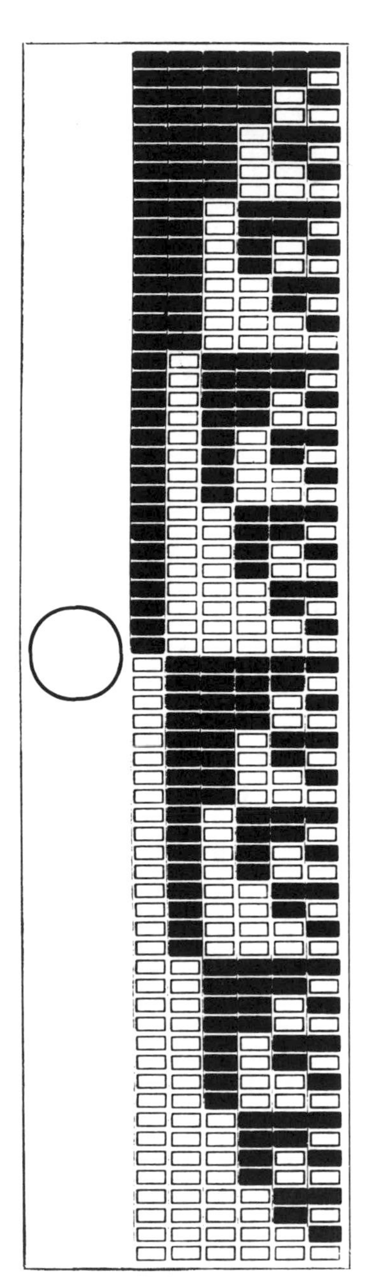

CHOW TSE'S DIAGRAM.

It will be of interest to compare Leibnitz's binary numbers with Chow-tsze's design; the similarity among which will appear as soon as 0 is identified with the black ▬ and 1 with the white ▭ spaces.

FIGURE 6. Facsimile of Page 228 of Carus's 1896 Article in *The Monist*.

Moritz Cantor declared Leibniz's interpretation mistaken and berated himself for having at first accepted it.[11] P. Carus, however observed that:

> Cantor seems to overlook that in this very respect the ancient Yang and Yin philosophy of the Chinese closely resembles Leibniz's idea, whether we regard the Kwa Figures of Fohy as numbers, or as a binary system of such symbols as are still more general and indefinite. The fact of both their presence and their philosophical significance remains the same and cannot be doubted.[12]

In further support of this point, Carus displayed a table of 6-bit strings and compared it with a Chinese diagram (Figure 3.6).

Raymond Clark Archibald called Leibniz's interpretation "worthless" and cited P. L. F. Philastre's monumental two volumes as the basis of his judgment. Philastre had been the first to translate *I Ching* into French. H. Brocard decided that even the oriental commentators had great difficulty with the nuances of the Chinese language in that ancient document; the task was, in his opinion, beyond any European.[13]

Nevertheless, combinatorial aspects susceptible to binary interpretations do exist in the Figures of Fohy. Blij, of the Mathematical Institute at Utrecht in the Netherlands, discusses these quite thoroughly in the 1967 article already cited.

NOTES TO CHAPTER III

[1]Heinrich Köhlern, *Lehr-Saetze des Gottfried Wilhelm von Leibniz*, (Leipzig: Joh. Meyers, 1720), p. 4.

[2]Rudolf August Nolte, *Leibniz Mathematischer Beweis der Erschaffung und Ordnung der Welt in einem Medallion an den Herrn Rudolf August*, (Leipzig: J. C. Langenheim, 1734).

[3]After a thorough search of the catalogs of applicable coin collections, including all known special Brunswickian collections, Dr. W. Jesse of the Städtisches Museum Braunschweig reported in his letter of November 2, 1965 that in his opinion, the proposed medallion had never been struck.

[4]Gottfried Wilhelm Leibniz, *Opera Omnia*, 6 vols. (Geneva: Fratres des Tournes, 1768), 3:352.

[5]C. I. Gerhardt, *Leibnizens mathematische Schriften*, 7 vols. (Berlin: A. Asher und Co., 1849), 3:656–669.

[6]Ibid., 4:92–103.

[7]Dietrich Mahnke, "Leibniz auf der Suche nach einer allgemeinen Primzahlgleichung," *Bibliotheca mathematica* 28 (May 1967):37.

[8]F. van der Blij, "Combinatorial Aspects of the Hexagrams in the Chinese Book of Changes," *Scripta Mathematica* 28 (May 1967):37.

[9]James M. G. Norwood, "I Ching," *Harper's Bazaar* 101 (October 1968):219.

[10]Wilhelm Ernst Tenzeln, *Curieuse Bibliothec*, (Franckfurt: Philipp Wilhelm Stock, 1705), pp. 81–112.

[11]Moritz Cantor, *Mathematische Beiträge zum Kulturleben der Völker*, (Hall: H. W. Schmidt, 1863), p. 49.

[12]P. Carus, "Chinese Philosophy," *The Monist* 6 (1896):229.

[13]H. Brocard, "Question 2960," *Intermédiaire des mathématiciens* 13 (1906):73.

·IV·

THE REST OF THE 1700S

SOME EARLY SCHOLARLY SEQUELS TO LEIBNIZ'S 1703 "EXPLICATION"

Serious reactions to the binary interpretation of the Figures of Fohy did not begin until 1863 with Cantor's (cited in previous chapter). This delay is not surprising; Europeans knew little about China during the 1700s and had to await significant advances in Sinology.

Only three early scholarly sequels to Leibniz's dyadics *per se* have been found. These are surprisingly few for a topic introduced with such enthusiasm by so prominent a man in the most prestigious scientific journal of its time, the *Memoires* of the Parisian Academy of Science. Moreover, none of the three appeared in these *Memoires*, where the carefully prepared and cross-referenced indices list nothing further on this topic at least through 1790.

The first of the three sequels, and the only one to appear before Leibniz's death in 1716, was an article by Petr Dangicourt published in 1710 in the first volume of *Miscellanea Berolinensia*.[1] This journal was published by the Berlin Academy of Science, which Leibniz had been instrumental in founding. The Latin title of the article is "*De periodis columnarum in serie numerorum progressionis Arithmeticae Dyadice expressorum*" and indicates that it deals with the analysis of columnar periods.

Unlike Leibniz, who had avoided the use of exponents in his "Explication," Dangicourt introduced binary notation by explaining that fedcba is to mean

$$f \cdot 2^5 + e \cdot 2^4 + d \cdot 2^3 + c \cdot 2^2 + b \cdot 2^1 + a \cdot 2^0,$$

where each letter represents either 0 or 1. To help explain the "carrying" in the operation of addition, he reminded his readers that

$$2^n + 2^n = 1 \cdot 2^{n+1}.$$

Dangicourt pointed out that if a column (in a vertical list of binary strings) has a period of length of 4, and if one marks every 3rd digit in that column, then the 12th digit will be a 'marked' digit, and also an 'end' digit of the columnar period. He saw that this involved the least common multiple of 3 and 4 being 12, and felt obliged to refer his readers to proposition 36 of the 7th book of Euclid, where a procedure is given for finding the least common multiple of two or more numbers. He found other excuses to refer to Euclid, but none to refer to Leibniz.

Dangicourt showed the following table of multiples of 8:

0000000	0
0001000	8
0010000	16
0011000	24
0100000	32
0101000	40
0110000	48
0111000	56
1000000	64
1001000	72
1010000	80
1011000	88
1100000	96
1101000	104
1110000	112
1111000	120

whose first three columns from the right are all zero. Nothing more startling than this appears in Dangicourt's 30-page article.

The second of the three sequels, Johan Bernard Wiedeburg's (1687–1766) postdoctoral thesis of 1718, has already been cited (p. 32) in reference to the binary medallion design. A facsimile of the full title page of this thesis appears in Figure 7.

Wiedeburg carefully reviewed Leibniz's "Explication," Fontenelle's "Nouvelle Arithmétique," and Dangicourt's "De periodis . . .". He seems to have been the first to publish a misreading of Fontenelle's account of Thomas Fantet de Lagny's work with dyadics. He reported falsely that Lagny had worked with logarithms expressed in binary notation, when in fact Lagny had *avoided* logarithms by a direct use of dyadics. It is possible that all subsequent, similarly false reports are based on Wiedeburg's misreading. But this is not probable, since his thesis has been less readily available than the *Histoire*, where Fontenelle's passage seemed to invite misreading by the rather unusual use of "natural logarithm" to mean "an aid to computation that avoids logarithms."

Wiedeburg followed up Leibniz's suggestion that dyadics can be used as the key to a minimum set of weights, these being the same as the binary place values. He showed in some detail how convenient such a system of weights would be, especially if the system of numeration were also a binary one. While he seemed enthusiastic about this in 1718, his 1725 book, which was a 1162-page introduction to mathematics, devoted no more than a few lines on page 11 to mentioning bases 2 and 4 as the inventions of Leibniz and Weigel respectively. Bishop Caramuel's 1670 work was still going unnoticed.

The third of the sequels is again a thesis. It was written by Johann Friedrich Weidler and appeared in 1719 in Wittenberg under the Latin title of *Dissertatio mathematica de praestantia arithmeticae decadicae, quatetractycam et dyadicam*. Weidler also carefully reviewed the work of Weigel, Leibniz, and Dangicourt on this topic, although he was apparently unaware of Wiedeburg's. He showed more extensive tables of equivalents and additional examples of the four fundamental operations in base 2 and base 4.

Q. D. B. V.

DISSERTATIO MATHEMATICA,

DE

PRÆSTANTIA ARITHMETICÆ BINARIÆ PRÆ DECIMALI,

QVÆ

RECTORE MAGNIFICENTISSIMO

SERENISSIMO PRINCIPE AC DOMINO

DN. GVILIELMO HENRICO,

DVCE SAXONIÆ, IVLIACI, CLIVIÆ, MONTIVM, ANGARIÆ, VVESTPHALIÆ, & RELIQVA,

DECRETV AMPLISSIMÆ FACVLTATIS PHILOSOPHICÆ

PRO LOCO

IN EADEM MORE MAIORVM RITE OBTINENDO

PVBLICÆ ERVDITORVM VENTILATIONI EXPOSITA EST

PRÆSIDE

IOHANNE BERNHARDO VVIDEBVRGIO,

MATHEMATVM PROFESSORE PVBL. ORD. & ALVMNORVM DVCALIVM INSPECTORE

ET RESPONDENTE

IOHANNE CHRISTOPHORO RHIEMIO, HILPERHVS. FRANC.

PHILOSOPH. & MEDICIN. CVLT.

H. L. Q. C.

AD D. XIII. APRIL. M DCC XVIII.

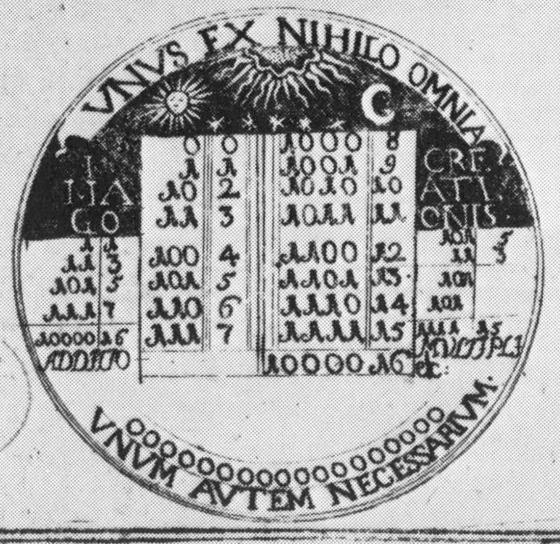

IENÆ, Literis KREBSIANIS.

FIGURE 7. Facsimile of the Title Page of the 1718 Postdoctoral Thesis Written by Johann Bernard Wiedeburg of Jena.

Charles XII and His Science Advisor Swedenborg

In his biography of Charles XII, of Sweden, Voltaire wrote:

> Some people have wanted to pass off this prince as a good mathematician; undoubtedly he had a deeply penetrating mind; but the proof usually given of his mathematical knowledge is not very conclusive; he would change the manner of counting by ten, and he would propose sixty-four in its place, because this number contained at once a cube as well as a square, and when divided by two, it would in the end reduce to unity. This idea could prove only that he loved the extraordinary and the difficult.

This passage appears on pages 342–343 of the 1823 edition, but is absent from the earlier Basel edition of 1781. Voltaire's own footnote to this passage stated:

> It proved also that he had penetrated, up to a certain point, the theory of numbers, until he knew the nature and the properties of these arithmetic scales.

A letter that Swedenborg (known as Swedberg before being raised to the nobility) had written to Goran Nordberg gives further details:

> Moreover, in the number 10, there is no cube, square, or biquadrate, the result being that with that number, great difficulties are met with in cubic and square reckonings. But if, in its place, we had made use of 8 or 16, then one would have had much greater advantage, and it would have carried with it greater ease, in that 8 is a cube of 2, and 16 a square of 4; so that it is at once accompanied by a cubic or square root. Moreover, by halving, this same number could be reduced to its principium or terminum primum, namely 1, without the intervention of any fraction; thus, 16, 8, 4, 2, 1. This same number could also be better fitted, and could better adapt itself, to the divisions in our coinage and measures, whereby many obscure and broken figures would be avoided.
>
> This, that is now brought forward in detail, was firmly held to by the King, and he desired that a trial be instituted with a number other than 10. And when, at this, it was said that such a trial could not be made unless new numbers and new names

should be invented and used, which must be wholly unlike those in common use, since were the least likeness retained it would at once result in bewilderment and confusion, this likewise was included in the trial.

By command the experiment was made with 8, which is a cube of 2, and which by halving stops exactly at its first number. And to this end new figures were invented and for each figure a new name. With these, the sequences were made according to the usual method, and it was applied to coins and measures and also to the cubic reckoning. When this trial was humbly laid before his Majesty, he did indeed think that all was satisfactory; yet it was seen that the Master wished rather a reckoning of greater extent and difficulty, wherewith he could have the opportunity the better to show his power and acumen of judgment and his deeper thought. To this end he raised the point as to whether there was not a number which contains within it both a cube and a square, and which likewise by halving could be brought to 1 without fractions. The number 64 was then suggested which is a square of 8 and a cube of 4, and which likewise could be halved to 1. We did not neglect, however, to point out that such a number would be too high, difficult, and almost impossible to work with. For if the turn should be first made at 64, and all prior thereto proceed in single numbers [digits], and from there a beginning be made with two numbers and when it has again come to the turning point 64, that is, has come to 64 times 64, being 4096, then to use three numbers; then in all reckoning with this system, and especially in multiplication and division, one would meet the difficulty of being obliged to hold in memory, a multiplication table consisting of 4096 numbers, instead of the one now in use, which consists of 80 to 90.

But the greater the difficulties occasioned therewith, the more brightly glowed the desire to make the trial and to show the possibility of a thing which we represented as requiring too much reflection to be brought into order and completion in a hurry. His Majesty then took it upon himself to work out and plan this method of counting. The next day he sent us his project, worked out on a whole sheet, with new numbers [digits] and new names. There he divided the aforesaid 64 figures into eight classes, and distinguished each class from the other by certain signs. On closer examination, these distinguishing signs were all found to be the initial and final letters and signs of the names that he used therein, but all in so pleasing and comprehensive a way that, after knowing the first 8 numbers, one had no difficulty in learning the rest of the sequence to 64. By the side of each and every number were set new names, and, on the 8 numbers in the first class, names such as could at once be remembered, while the remaining numbers were marked with a differentiation such that, without the difficulty of burdening the memory, one could easily recite all the numbers in order, and the counting could be done according to rule. And when one had reached 64, and would continue to 64 times 64 or 4096, with three figures, then followed, by means of the abovementioned differentiations, new names, in such order, one after the other, all with a natural and self-demonstrating variation, that one had new and fitting names for numbers as high as could be desired, so that there could be no number so high as not always to carry with it a change in name, according to the principle and rule formulated. As already stated, this same project which his Majesty set up with his own hand, and which I still have in my possession in the original was sent to me that, with its guidance, I might form a table which would show the difference, in respect both to names and to numbers, between this and ordinary method of counting.

On this same project it was also shown how both multiplication and division should be done, the intention therewith being to meet the great difficulties.

When by such high command, the said project was to be further expanded, and the opportunity was given me for closer reflection as to whether this reckoning could be set up in a better and more convenient way, then, however much I worked on it, I could yet find nothing that could be improved on, nor in my humble opinion, could any other man have found anything, even though he used his time on mathematics alone.[2]

A footnote by translator Alfred Acton indicates that the written proposal from Charles XII is not available. The only knowledge comes from Swedenborg's letter quoted above.

In his *A New System of Reckoning*, Swedenborg discusses only the base eight, or octal system and uses the eight digits

o l s n m t f v

for this purpose. It is noteworthy both that he gave complete addition and multiplication tables for his octal system, and that he adopted Napier's bones to it.

The "Translator's Preface" reveals that Swedenborg's own brother-in-law refused to publish this work on the grounds that it was too revolutionary and might result in a suspicion among the people of Sweden that it was the precursor of a coming change in coinage—with which they had already had bitter experience. Other print shops could not cope with the special demands involved, so Swedenborg prepared an especially beautiful manuscript version for presentation to the King. Unfortunately Charles was hit and killed by a cannon ball on November 30, 1718 before the presentation could take place. According to this translator, Swedenborg anonymously published a little tract in November 1719 entitled, *A Suggestion for so dividing our Coinage and Measures that Calculations can be facilitated, and all Fractions avoided*. This work advocated the adoption of the *decimal* system in Swedish coinage, weights, and measures.

It is not clear whether either Swedenborg or his King was ever *serious* about introducing a nondecimal base for common use, but it is clear that among nondecimal bases the two men preferred a number that is a power of two, such as 8, 16, 32, or 64. Some writers came to believe that Charles XII favored base 12. Perhaps they had simply distrusted reports about a base so high as 64, or they had equated *antidecimal* with *duodecimal*. Levi Conant, for example, wrote:

> So palpable are the advantages of 12 from this point of view that some writers have gone so far as to advocate the entire abolition of the decimal system and the substitution of a duodecimal system in its place. Charles XII, of Sweden, may be mentioned as an especially zealous advocate of this change, which he is said to have had in actual contemplation for his own dominions at the time of his death.[3]

Conant's prominence has assured wide circulation of this error.

Euler's and Beguelin's Use of Dyadics

Christian Goldbach, in a letter dated December 1, 1729, called Leonard Euler's attention to the Fermat Conjecture:

Every number of the type $F_n = 2^k + 1$ (*where* $k = 2^n$) *is a prime.*

The first few of these Fermat numbers, namely 3, 5, 17, 257, and 65537 (for n = 0, 1, 2, 3, and 4) are indeed prime. Fermat admitted in a 1654 letter to Pascal that he had no proof of that conjecture.[4] Even the next Fermat number (F_5 = 4,294,967,297), not to speak of those following, is rather large and its primeness difficult to verify or disprove. Thus, in 1729, at the time of the Goldbach letter, still no proof or disproof of Fermat's Conjecture was known.

In binary notation, numbers of the type $2^q + 1$ take on a particularly simple form, namely 1000...0001, i.e., a binary string that begins and ends in the digit 1 with all other digits 0, thus

$$F_5 = 2^{32} + 1 = 100000000000000000000000000000001$$

where the expression $(2^{32}_s + 1)$ itself may be interpreted as being in binary notation or the expanded form thereof.

According to Gustaf Enestrom, Euler submitted a counterexample to Fermat's Conjecture to the Academy at Petersburg on September 26, 1732.[5] Using expanded binary notation, Euler showed that when $(1 + 2^7 + 2^9)$ is divided into $(2^{32} + 1)$ the result is the whole number

$$(1 + 2^7 + 2^8 + 2^{10} + 2^{11} + 2^{12} + 2^{13} + 2^{17} + 2^{18} + 2^{21} + 2^{22}).$$

Like Dangicourt, Euler pointed to the identity

$$2^n + 2^n = 2^{n+1}$$

as the key to the binary arithmetic involved.

N. Beguelin, apparently unaware of Euler's work on this topic, gave similar results in 1777, using his own abbreviation for expanded binary notation.[6] Beguelin described

$$(2^{32} + 1) = (0.7.9)(0.7.8.10.11.12.13.17.18.21.22)$$

where (0.7.9) stands for $(2^0 + 2^7 + 2^9)$. Translated into decimal notation, this reads:

$$4{,}294{,}967{,}297 = (641)(6{,}700{,}417)$$

MISCELLANEOUS WRITERS OF THE MID 1700s.

In 1740 Christoph Friedrich Vellnagel treated bases 2, 3, 4, 5, 6, 7, 8, 9, and 12, showing addition and multiplication for the first two of these in his 36-page booklet.[7] It is noteworthy for showing arithmetic operations in base 3, but not otherwise.

In 1746 Francesco Brunetti was the first to publish serious work on binary fractions.[8] His 55-page booklet devotes the first 38 pages to whole numbers; he then introduced some fractions. An example of the latter is 191.45, whose binary equivalent is $10111111\frac{1001}{10100}$, which simply involves the fraction 9/20 with numerator and denominator in binary notation. True binary fractions (those whose denominators are a power of two) are introduced on page 41 by showing:

$$\begin{aligned} 101.1 &= 5\tfrac{1}{2} \\ 111.11 &= 7\tfrac{3}{4} \\ 0.001 &= \tfrac{1}{8} \\ 1110.0101 &= 14\tfrac{5}{16} \end{aligned}$$

Brunetti pointed out that $2/3 = 0.1010101010\overline{10}\ldots$, but in a later dyadic computation he seemed satisfied to use 0.10101 for 2/3, i.e., he substituted 21/32 for 2/3. He gave examples of multiplication and division involving dyadic fractions, of which the following is one:

$$\begin{array}{r} 1.011 \\ 1.1 \\ \hline 1011 \\ 1011 \\ \hline 10.0001 \end{array}$$

In effect he pointed out that the infinite dyadic fraction $0.101010\overline{10}\ldots$ is the geometric series

$$1/2 + 1/8 + 1/32 + 1/128 + 1/512 + \ldots$$

and that $1.0111111\overline{1}\ldots$ stands for

$$1 + 1/4 + 1/8 + 1/16 + 1/64 + \ldots .$$

In a section devoted to "Dyadic Logarithms," Brunetti displayed:

1	0.00000000000000000000
10	1.00000000000000000000
11	
100	10.00000000000000000000
101	
110	
111	
1000	11.00000000000000000000

where the left column contains the "natural numbers" and the right column their logarithms to the base 2. Every entry is in dyadic notation. Aware that he had omitted nonintegral logarithms, Brunetti somewhat gingerly discussed that if one had found log(11), for example, then twice that would give the missing dyadic value for log(110). He gave a table of decimal values of powers of two up to 2^{40}, but those after 2^{34} = 17,179,869,184 are in error.

A year later, in 1747, Johann Berckenkamp's 59-page booklet appeared.[9] In addition to all bases below 10, Berckenkamp dealt rather superficially with bases 12, 13, 15, 24, and 30; he restricted himself to whole numbers and avoided nondecimal arithmetic operations. However he provided such "theorems" as, for example, his Theorem 57:

$$1221_3 = 52_{10}$$

He used 0, a, b, ..., x, y, and z (omitting j, o, and u) as the 24 digits for base 24, where

$$a00 = 576$$

and where, according to his "Theorem 89,"

$$xyxv = 331{,}167.$$

He used the same 24 digits, augmented by α, β, γ, δ, ε, and ζ, for base 30, where

$$a\alpha = 54.$$

Etienne Bezout[10] explained, in substance, apparently as early as 1764, that for an arbitrary base β, having 0, 1, 2, ... , $(\beta - 1)$ as its β digits, the string of digits *bcdef* represents

$$b\beta^4 + c\beta^3 + d\beta^2 + e\beta + f$$

which, in the case of $\beta = 12$, would mean that the string 57643 represents $5 \cdot 12^4 + 7 \cdot 12^3 + 6 \cdot 12^2 + 4 \cdot 12 + 3$. Moreover, Bezout gave, in substance, the following procedure for finding the base β representation of a number N:

If β^p is the greatest power of β contained in N, then divide N by this β^p and note the quotient a and the remainder r. This number r should now be divided by β^{p-1} calling the resulting quotient b and the remainder s. Continuing in the same manner until a remainder of k is obtained, such that k is zero or less than β, then *abcd...k* will be the base β representation of the number N.

To illustrate this, suppose the duodecimal equivalent of 12891 is to be found. Divide 1728 ($=12^3$) into 12891, obtaining a quotient of 7 and a remainder of 795. Now divide 795 by 144 getting a quotient of 5 and a remainder of 75. Now divide 75 by 12, obtaining a quotient of 6 and a remainder of 3. It now follows that $1289_{10} = 7563_{12}$.

Thus, in a page and a half, Bezout managed to explain all bases β at once, illustrating his general principles by some examples for base 12. He thus accomplished more in his page-and-a-half than Berckenkamp had in his entire 59 pages.

The eight pages that Georges Buffon devoted to nondecimal

numeration are of similar high quality.[11] He left no doubt of his preference for base 12 among all possible bases, mostly because base 12 numeration would dovetail better with many already established measures and their divisions (such as the year and its twelve months) and because the number 12 was rich in divisors. He saw similar advantages in base 60, but considered this out of the question because of the 50 additional symbols that would be required. Buffon's explanation of base systems in general and a conversion procedure from base 10 to an arbitrary base was similar to Bezout's.

Buffon went beyond Bezout in the following respects:

(i) He took the slightly more sophisticated view that the last remainder k (in Bezout's conversion procedure) is also the last quotient, the last division being by β^0, i.e., by 1.

(ii) He showed that the highest power of β, namely β^p, that is contained in N, can be found by noting that $N = \beta^v$ yields $v = (\log N)/(\log \beta)$ and that the integral part of v must be equal to p. So, for example, when $N = 1738$ and $\beta = 5$, then $v = (\log 1738)/(\log 5) = 4^+$ and hence $p = 4$. It follows that the base 5 representation of 1738 must have 5 digits, as in fact it does, since $1738 = 23{,}423_5$.

Georg Brander's booklet of 40 pages appeared in 1775.[12] Although its author was a member of the Bavarian Academy of Science, the book was written in a light conversational German. In the preface, Brander expressed the hope that this work on dyadics might serve the reader for recreational purposes, and perhaps even for keeping one's personal accounts unintelligible to prying eyes. He indicated that his knowledge of this topic had come from Wiedeburg's dissertation. From the same source came his belief that the ancient Chinese had known and then lost the art of dyadics, thereby giving the great Leibniz an opportunity to rediscover it.

Brander restricted his work to whole numbers, except for an occasional use of fractions like 1/1001 (meaning 1/9). He showed numerous examples of the four fundamental operations as well as the pulling of square and cube roots. However, all his examples and discussion involved only the base two.

Whereas the Bezout-Buffon procedure for converting a whole number N from decimal to β-adic notation had called for division by successively smaller powers of β (starting with the highest contained in N), Brander's procedure called for repeated division by 2. For example, Brander found the binary equivalent of 144, as follows:

	144 is even,	hence one writes	0
halved	72 is even,	hence one writes	0
halved	36 is even,	"	0
halved	18 is even,	"	0
halved	9 is odd,	"	1
halved	4 is even,	"	0
halved	2 is even,	"	0
halved	1 is odd,	"	1

Starting from the bottom the digits on the right are now to be written next to each other, thusly:

10010000

which is the binary equivalent of 144.

This procedure exposed the binary digits in the opposite order. Brander offered no explanation as to why his procedure would give the desired result.

Felkel's 1785 Paper on the Periods of β-Adic Fractions

For the fraction 1/7, whose decimal equivalent, $0.\overline{142857}\ldots$ has a period of 6 digits, Anton Felkel showed the following base β equivalents:[13]

For

$\beta = 3$	$0.\overline{010212}\ldots$	
$\beta = 5$	$0.\overline{032412}\ldots$	complete
$\beta = 17$	$0.\overline{274e9c}\ldots$	periods
$\beta = 2$	$0.\overline{001}\ldots$	
$\beta = 4$	$0.\overline{021}\ldots$	1/2 complete
$\beta = 9$	$0.\overline{125}\ldots$	periods
$\beta = 11$	$0.\overline{163}\ldots$	
$\beta = 6$	$0.\overline{05}\ldots$	1/3 complete
$\beta = 13$	$0.\overline{1b}\ldots$	periods
$\beta = 8$	$0.\overline{1}\ldots$	period of length 1

where the letters a, b, c, etc. have been introduced as digits for ten, eleven, twelve, etc.

As Felkel explained in his introductory remarks, since for a fraction of the type 1/p where p is prime, the maximum length is (p − 1); he decided to call periods of that length *complete* and the shorter ones *incomplete*. He decided to distinguish between periods of odd and even lengths, since he had concluded, in substance, that

Theorem 4.1: *The digits of the second half of a period (in the β-adic expansion of a fraction 1/p) of* even *length are the (β−1) complements of the corresponding digits of the first half.*

For $\beta = 10$ and $p = 7$, for example, this means that if the period 142857 is separated into two halves and the second half is written under the first, subsequent addition of these two halves will result in all 9's, thusly:

$$\begin{array}{r} 142 \\ \underline{857} \\ 999 \end{array}$$

For $\beta = 5$ and $p = 7$, the period 032412 will give 4 as the sum of corresponding digits from the first and second half.

Felkel made use of Theorem 4.1 by abbreviating, for example:

$$0.\overline{01020413321\ 43424031123}\ldots(1/23 \text{ in base } 5)$$

to $\quad 0.01020413321\ \mathbf{I}$,

where the symbol **I** was to indicate that only the first half of the period had been given, but that the second half could easily be reconstructed since its digits would be the 4's complements of the corresponding members of the first half.

The β-adic equivalents of 1/7, displayed earlier, show that 1/7 may have periods of length 1, 2, 3, and 6, precisely the divisors of 6. Felkel noted, though in this case did not display, that for 1/211 all 16 divisors of 210 will appear as period lengths. He indicated that 13 of these can be found (in decreasing order) by converting 1/211 to the bases 2, 4, 18, 26, 5, 10, 19, 12, 58, 15, 55, 14, and 210. He announced, in substance:

Theorem 4.2: *If* p *is prime, then the divisors of* (p − 1) *are the only numbers that can appear as period lengths of the β-adic representation of* 1/p; *moreover, every such divisor will appear as a period length.*

In an apparent effort to balance Theorem 4.1 with another theorem about periods of *odd* length, he stated, in effect:

Theorem 4.3: *If the β-adic representation of* 1/p *has a period of* odd *length, then the fraction* (p − 1)/p (*which is the 1's complement of* 1/p) *has a period of the* same *length, with each digit being the* (β − 1)*'s complement of the corresponding digit of the period of* 1/p.

Felkel failed to notice, or at least to indicate, that the hypothesis of this theorem could be weakened to "a period of odd *or even* length."

Felkel pointed out that in order to find the β-adic equivalent of any rational fraction, for any β, one should proceed as one does in base 10, i.e., divide the denominator into the numerator and do so in base β arithmetic. Thus, to get the dyadic equivalent of 1/13, one should divide 1101 into 1, or to get the base 30 equivalent of that fraction, d into 1, where d is the digit for thirteen. In the latter example, as for all large β, Felkel showed that the inconvenience of learning new digits and multiplication facts may be circumvented by expressing the larger digits decimally. Thus instead of writing

$$1/13 = 1/\mathrm{d} = \mathit{0.296slo},$$

one could write

$$1/13 = \quad 0.\boxed{2}\boxed{9}\boxed{6}\boxed{27}\boxed{20}\boxed{23}$$

where $\boxed{27}$, for example, has been boxed in since it represents the *single* digit s. (His assignments start a = 10, b = 11, etc., but he omitted the letter j, and hence *s* = 27, etc.) He reminded his readers that for $\beta = 10$, the usual procedure exposes successive digits of the decimal equivalent by repeatedly "attaching a zero" and that this amounts to successive multiplication of the dividend by 10, or in general by β.

He further pointed out that $1/9 = 0.\overline{1}\ldots$ and $1/11 = 0.\overline{09}\ldots$ are special cases of

Theorem 4.4: $1/(\beta - 1) = (0.\overline{1}\ldots)_\beta$

and

Theorem 4.5: $1/(\beta + 1) = 0.\overline{0\mathrm{Z}}\ldots)_\beta$ *where* $\mathrm{Z} = \beta - 1$.

Each of these theorems reflects an algebraic identity.

A century before Felkel, Jordaine had shown how to convert a terminating decimal fraction into a duodecimal one. In effect, Felkel modified Jordaine's procedure to the point where, in Leonard Dickson's words, he could be credited with having shown "how to convert directly a period fraction written to one base into one to another base."[14]

Two examples from Felkel's paper will illustrate the procedure.

Example 1.

a. 0.076923
b. 0.461538
c. 2.7692~~28~~ (30)
d. 4.61538~~0~~ (4)
e. 3.69230~~4~~ (7)
f. 4.15384~~2~~ (6)
g. 0.923076
h. 5.5384~~56~~ (61)
i. 3.23076~~6~~ (9)
k. 1.38461~~4~~ (5)
l. 2.30769~~0~~ (2)
m. 1.84615~~2~~ (3)
n. 5.0769~~18~~ (23)

This shows how he found $(0.\overline{076923}\ldots)_{10}$ to equal $(0.\overline{024340531215}\ldots)_6$. Seemingly, he worked only with a single period, 076923, which he repeatedly multiplied by 6 to expose, one by one, the digits of the base 6 equivalent, subtracting after each such multiplication the newly exposed digit. In substance, however, he was multiplying not just a single period, but $0.076923\ 076923\ \overline{076923}\ldots$. He took this into account when he crossed out (in step c.) the digits 2 and 8 and substituted the digits 3 and 0, since he was "carrying" a 2 from the multiplication of the 2nd period. Similarly, at each step he showed how he modified his product because of the "carry." For each step, the amount carried (from one period to the next) is, of course, equal to the newly exposed base 6 digit. The integral parts of the fractions listed vertically give the digits of the desired base 6 equivalent. In Example 2, he shows how the given $(0.\overline{032412}\ldots)_5$ was converted to $(0.\overline{010212}\ldots)_3$.

Example 2

a.	0.203241	
b.	1.120323̸	(4)
c.	0.412032	
d.	2.241201̸	(3)
e.	1.32411̸4̸	(20)
f.	2.032410̸	(2)

where the given period, 203241, was repeatedly multiplied by 3, using base 5 arithmetic. The product was appropriately increased for any "carry" and decreased by the newly exposed digit before the next multiplication.

Felkel pointed out (without giving any examples) that his procedure could be modified for use with irrational numbers. If one desired the first m digits of the base γ equivalent of an irrational number x, one would have to use the first n digits of the given base β equivalent of x, making sure to pick n large enough to have included all digits that might influence the desired string of digits.

Duodecimal versus a Decimal Metric System

On October 27, 1790, the Metric Commission, consisting of Jean Charles Borda, Joseph Louis Lagrange, Antoine Laurent Lavoisier, Mathieu Tillet, and Marie Jean Antoine Nicolas Caritat Condorcet, made a "Rapport" to the Academy of Science in Paris, which indicated that the Commission had briefly considered proposing the replacement of decimal arithmetic for common use by the duodecimal system and a metric system duodecimalized to match. However, it was decided that proposing a change in common arithmetic, in addition to a change in weights and measures, would doom their project to failure. While it was important that both arithmetic and the metric system have the same radix, this could be accomplished by leaving common arithmetic intact and decimalizing the system of weights and measures.

Any advantages of base 12 would be due to its richness in divisors relative to 10. Also, certain common fractions such as 1/3 and 1/4 would have simpler equivalents in base 12 notation. As one can gather from a later paper by Jean Delambre at least one member of the Commission, namely Lagrange, refused to concede even such a slight theoretical advantage to base 12.[15] He argued that poorness in divisors was an advantage, and that perhaps they should consider a prime number, such as 11. No proper fraction (with 11 as a denominator) would then be reducible and each would neatly preserve 11 as a denominator.

To this day, lovers of the duodecimal numeration system have been fighting a rearguard action against wider adoption of the decimalized metric system.

Lagrange, Laplace, Lamarque, and Legendre

In his first letter to future teachers at the École Normale in Paris in 1795, Joseph Lagrange said:

> Arithmetic is divided into two parts. The first is based on the decimal system of notation and on the manner of arranging numerals to express numbers. The first comprises the four common operations of addition, subtraction, multiplication, and division,—operations which, as you know, would be different if a different system were adopted, but, which it would not be difficult to transform from one system to another, if a change of systems were desirable.
>
> The second is independent of the system of numeration. It is based on the consideration of quantities and on the general properties of numbers.[16]

Lagrange further stated, in substance:

Theorem 4.6: *A number is divisible by* $(\beta - 1)$ *if the sum of its* β*-adic digits is divisible by* $(\beta - 1)$. This is a special case of Pascal's General Divisibility Theorem, though Pascal had fallen short of actually stating it.

Lagrange made a slight improvement on Pascal's Divisibility Test for 7 ($\beta = 10$), by making use of negative remainders. Thus, Pascal's Theorem 2.5 had involved the remainders 1, 3, 2, 6, 4, and 5, but Lagrange used 1, 3, 2, -1, -2, and -3.[17]

With these future teachers, Lagrange did not discuss his supposed preference for a prime base, such as 11. His colleague Pierre-Simon Laplace, who lectured at the École Normale the same year, let it be known that the most preferable of all possible bases would be one that is not too large and that has a great many divisors. Base 12 met these requirements. Laplace indicated further that only two additional digits would be needed, but the Commission on the metric system rejected 12 in favor of 10, lest the entire project be jeopardized. To convert a number from decimal to duodecimal notation, Laplace recommended repeatedly dividing by 12 (or by β, if one were converting to β-adic notation) and noting the successive remainders, which, in reverse order, would constitute the digits in the new notation. Before Laplace stated his preference for $\beta = 12$, he reviewed Leibniz's work on $\beta = 2$. He accepted without reservations the idea that the ancient Chinese had used dyadics and that Leibniz had rediscovered it, although he was bemused by Leibniz's having seen dyadics as the image of Creation.

The year 1795 was the second year of the Republic during the French Revolution, which explains in part why such outstanding mathematicians as Lagrange and Laplace were lecturing at a teacher's college. During the third year of the Republic, as the title page proudly proclaimed, the second edition of Jean Lamarck's three-volume work on French flowers appeared. The first edition had been published in 1778. It is mentioned here only because Giuseppe Peano was later to declare that Lamarck's classification scheme was an application of the binary system. The scheme is more often referred to as being dichotomous, and there is no obvious indication that Lamarck himself thought he was applying the binary system.

Before commenting on a minor contribution to this topic by

Adrien-Marie Legendre, it will be convenient to introduce the concept of γ-adic coded β-adic notation. Felkel had called attention to the fact that for a large β it may be convenient to sidestep the introduction of additional symbols and use instead decimal notation to represent the digits. Thus he had shown a boxed in $\boxed{27}$ instead of the digit *s*. The definition of a standard numeration system does not specify the nature of the β symbols that might be used, and hence both 0.296*slo* and $0.\boxed{2}\boxed{9}\boxed{6}\boxed{27}\boxed{20}\boxed{23}$ are in *standard* base 30 notation. The latter, however, since it also involves base 10, may be called decimal coded base 30 notation. Similarly, Caramuel's use of base 60 involved the decimal coded base 60 system. The more sophisticated versions of *γ-adic coded β-adic notation* have leading zeros, so that every base β digit will be coded into a γ-adic string of equal length. If Felkel's version were so modified, it would appear as

$$\boxed{00}.\boxed{02}\boxed{09}\boxed{06}\boxed{27}\boxed{20}\boxed{23}$$

or simply as

$$00.02090626272023$$

where the reader carries the burden of counting off to decide where one base 30 digit ends and the next one begins.

In a mere footnote, Legendre gave a hint as to how one might convert quickly from decimal to binary notation when the given number N is rather large.[18] For N = 11183445, to use his example, he divided N by 64, obtaining a remainder of 21 and 174741; then dividing that quotient by 64, he obtained a remainder of 21 and a quotient of 2730. Upon the next division by 64, he obtained a remainder of 42 and a quotient of 42. In effect, he had converted N to decimal coded base 64, i.e.,

$$N = (42,42,21,21)_{64}$$

and now noting that $42 = 101010_2$ and $21 = 10101_2$, he wrote

$$N = 101010\ 101010\ 010101\ 010101.$$

This last string of binary digits (especially with the grouping shown) may be interpreted as being in binary coded base 64 notation (note some leading zeros supplied by Legendre), but this is at the same time true binary notation. Underlying this is a principle tacitly assumed by Legendre, namely

Theorem 4.7: *If* $\beta = \gamma^i$ *for some positive integer* i, *and* $N = (a_n \ldots a_o)_\gamma$ *is a* γ*-adic string of* (n + 1) *digits where* (n + 1) *is a multiple of* i, *if necessary by the introduction of leading zeros; if these* (n + 1) *digits are grouped into* (q + 1) *groups of* i *digits each, calling these (starting from the right)*, G_0, G_1, ... G_q, *then*

$$(a_n \ldots a_0)_\gamma = (G_q \ldots G_0)_\beta.$$

The right side of this last equation may be called a *γ-adic coded β-adic string.*

A frequent, tacit application of this theorem occurs when the decimal string 1984 is read "nineteen hundred and eighty-four," i.e., $(19,84)_{100}$, or as a decimal coded base 100 string.

Why was Legendre interested in the binary equivalent of certain large numbers? He was introducing certain theorems concerning the "Legendre symbol," with which he could settle whether or not a prime p was a divisor of the expression $(x^2 + a)$ for a given positive integer a and for some integer x. In his Example I, he settled in the negative, whether the prime p = 1013 is a divisor of $(x^2 + 601)$. Without his newly introduced theorems, Euler's criterion would have had to suffice. This called for raising the integer a to the kth power where $k = (p - 1)/2$, and discarding multiples of p along the way. A final outcome of (−1) would settle the question in the negative, (+1) in the affirmative, no other outcome being possible.

Thus, to verify his result using older methods, the number 601 would have to be raised to the 506th power, discarding multiples of 1013 along the way. This older method could be expedited by noting that

$$\begin{aligned} 506 &= (111\ 111\ 010)_2 \\ &= 2^8 + 2^7 + 2^6 + 2^5 + 2^4 + 2^3 + 2^1 \\ &= 256 + 128 + 64 + 32 + 16 + 8 + 2. \end{aligned}$$

Thus 601 would have to be squared 8 times to find 601^{256}, 7 times to find 601^{128}, and so on. The product of these individual results would give the final result for 601^{506}. In Legendre's Example III, similar verification by older methods involved raising 1459 to the $11{,}188{,}445^{th}$ power, hence the desirability of knowing the binary equivalent of the number 11,188,445.

NOTES TO CHAPTER IV

[1]Petr Dangicourt, "De periodis columnarum in serie numerorum progressionis arithmeticae dyadice expressorum," *Miscellanea Berolinensia* 1 (1710): 336–376.

[2]Emanuel Swedborg, *A New System of Reckoning Which Turns at 8*, (Philadelphia: Swedenborg Scientific Association, 1941), pp. 7–8.

[3]Levi Leonard Conant, "Primitive Number Systems," *Smithsonian Institution Annual Report* (1892):588.

[4]A. P. Juskevic and E. Winter, *Leonard Euler und Christian Goldbach*, (Berlin: Akademie-Verlag, 1965), p. 29.

[5]See both Gustaf Eneström, *Verzeichnis der Schriften Leonhard Eulers*, (Leipzig: B. G. Teubner, 1910), p. 7, and Leonard Euler, *Opera postuma*, (Petropoli: Fuss und Fuss, 1892), 1:169–171.

[6]N. Beguelin, "Application de l'Algorithme exponentiel a la recherche des facteurs des nombres de la forme $2^n + 1$," *Nouveaux Memoires de l'Academie Royale des Sciences et Belles-Lettres*, (*Berlin*) (1777):251.

[7]Christoph F. Vellnagel, *Numerandi methodi*, (Jena: Franciscus, 1740).

[8]Francisco Saverio Brunetti, *Arimmetica binomica e diadica*, (Rome: Bernabo e Lazzarini, 1746).

[9]Johann Albert Berckenkampf, *Leges numerandi quibus numeratio decadia Leibnizia dyadica*, (Lemgovia: 1747).

[10]The author examined a copy of Etienne Bezout, *Course de mathématique*, pp. 19–20 that lacked a title page. A reference to a 1779 publication would suggest a publication date of 1780 or later. The author's preface indicated two earlier versions of this text, one slanted for the French Navy, the other for the Artillery. The present text represents an integration of the two. Lewis Carl Seelbach, "Duodecimal Bibliography" indicates that the duodecimal system was similarly treated by Bezout in these earlier versions, one of them being dated 1764.

[11]Georges Louis Leclerc Buffon, *Oeuvres complètes*, 12 vols. (Paris: Garnier Frères, 1938), 12:488–495.

[12]George Friedrich Brander, *Arithmetica binaria sive dyadica das ist Die* Kunst nur mit zwey Zahlen in allen vorkommenden Fällen sicher und leicht zu rechnen, (Augsburg: the Author, 1775).

[13]Anton Felkel, "Verwandlung der Bruchsperioden, nach den Gesetzen verschiedener Zahlensystems," *Ceska spolecnost Nauk, Prague Abhandlungen*, series 2 (1785):135–174.

[14]Leonard Eugene Dickson, *History of the Theory of Numbers*, 3 vols. (New York: G. E. Steckert and Company, 1934), 1:161.

[15]Jean Baptiste Joseph Delambre, "Notice sur la vie et les ouvrages de M. Lagrange," *Memoires de la classe des sciences mathématiques et physiques de l'institute Impérial de France* 13 (1812):LXVI.

[16]Joseph Louis Lagrange, *Lectures on Elementary Mathematics*, trans. Thomas J. McCormack (Chicago: The Open Court Publishing Company, 1898), p. 1.

[17]Ibid., p. 33.

[18]Adrien-Marie Legendre, *Essai sur la théorie des nombres*, (Paris: Duprat, 1798), p. 229.

·V·

THE NINETEENTH CENTURY

I. THE FIRST QUARTER

Gauss's *Disquisitiones Arithmeticae*

Carl Gauss's *Disquisitiones Arithmeticae* appeared in 1801. It dismissed nondecimal numeration with a mere footnote. "For brevity we will restrict the following discussion to the system which is commonly called decimal, but it can easily be extended to any other."[1] This footnote referred to paragraph 312, in which the periods of the decimal equivalents of rational fractions were being treated.

Gauss's book introduced the theory of modular congruences, and thus provided the tools for simple proofs of the theorems announced by Felkel and the divisibility rules of Pascal. This was certainly not in line with the expectations Leibniz had a century earlier when he introduced the binary system as a promising theoretical tool. As Mahnke was later to see it, the historic development went differently from Leibniz's hopes.[2] Instead of the study of periodic decimal and nondecimal fractions contributing to the development of number theory, it was to be the systematic construction of number theory by Gauss that became the key to a full understanding of periodic decimal and nondecimal fractions.

Ozanam's *Recreations in Mathematics*

The 1803 edition of Jacques Ozanam's *Recreations in Mathematics* (enlarged by Montucla and translated into English by Charles Hutton) devoted pages 2–9 of volume I to nondecimal numeration. Ozanam explained Leibniz's binary system and then commented on the binary interpretation of the Figures of Fohy. "It is very singular, that a Chinese enigma should find its Oedipus only in Europe, but perhaps in this explanation there is more ingenuity than truth." Ozanam clearly considered base 12 the best possible base; he cited all the usual reasons of richness in divisors and dovetailing with certain existing units of measures and their subdivisions. In fact, he chided Simon Stevin for having sought such dovetailing by decimalizing weights and measures, rather than by duodecimalizing the arithmetic.

Peter Barlow (1810–14)

Peter Barlow dealt with nondecimal numeration in three separate publications during 1810–14. In 1810 he published an article entitled "On the Method of Transforming a Number from one Scale of Notation to Another, and its Application to the Rule of Duodecimals." The article did not consider 10 the best of possible bases, but accepted it as satisfactory and certain to remain in common use. As to how to transform a number from decimal to duodecimal notation, he recommended the method suggested by the following example, where it is required that decimal 1728 be transformed to the "duodenary scale." Here A = 10 and B = 11.

10	=	A			
10^2	=	A×A	=	84	
10^3	=	84×A	=	6B4	
10^4	=	6B4×A	=	5954	etc.
1728	=	$10^3 + 10^2 \times 7 + 10 \times 2 + 8$			
8	=		=	8	
10×2	=	A×2	=	18	
$10^2 \times 7$	=	84×7	=	4A4	
$10^3 \times 1$	=	6B4×1	=	6B4	
1728_{10}	=		=	1000_{12}	

Barlow gave three examples of "applications to the rule of duodecimals," one of which involved the requirement that 17 feet, 3′, 4″ (meaning 3 inches and 4/12 inches) be multiplied by 19 feet, 5′, 11″. He recommended transforming to 15.34 17.5B with the resulting answer of 240.9688_{12} = 336 ft, 9′, 6″, 8‴, 8⁗.

Barlow's 1811 *An Elementary Investigation of the Theory of Numbers* contained a chapter "On the different Scales of Notation and their Application to the Solution of Arithmetical Problems." Here he recommended repeated division by β as the best method of transforming a number from decimal to β-adic notation. There is no indication that Barlow was familiar with Gauss's theory of congruences; he most certainly avoided the use of congruence notation both in statements and proofs of his theorems. Nevertheless, he advanced beyond the Pascal type of divisibility theorem by giving not only a criterion for divisibility by K, but also a criterion for predicting precisely what the remainder will be in case the number is not divisible by K. Thus he stated in substance:

Theorem 5.1: *If* T_s *and* T_a *are the simple and alternating* β*-adic digit sums of* N, *then* N *will have the same remainder as* T_s *upon division by* $(\beta - 1)$, *and* N *will have the same remainder as* T_a *upon division by* $(\beta + 1)$.

For base 10, this theorem is the basis of the checking procedure known as "casting out nines," and also for a similar procedure involving eleven. Barlow recommended the latter in addition to the former whenever one wanted additional assurances of the accuracy of arithmetic results.

Barlow explained, as had some earlier writers, that base 2 notation was a key to the fact that every integral number of pounds could be balanced on a scale by having weights on the other restricted entirely to single specimens of each binary place value. He failed to show, or even mention, that such a set of weights would constitute a *minimum* set of weights.

The fact that

$$716 = (222112)_3$$

suggests that an object weighing 716 pounds could be balanced by using 2 weights of 3^5 each, 2 of 3^4, 2 of 3^3, 1 of 3^2, 1 of 3^1, and 2 of 3^0. More generally, the fact that a unique base 3 string exists for each positive integer readily suggests that a set of weights consisting of *two* specimens of 3^i pound weight (for each integer from $i = 0$ to $i = k$) would suffice for weighing any object of an integral number of pounds up to $3^{k+1} - 1$.

Barlow showed that a smaller set of weights, consisting of but a *single* specimen of each 3^i pound weight, would also suffice, although to not quite as high a limit, if one were permitted to place weights in *either* pan. Thus a 716 pound object could be weighed by placing a 3^2, a 3^1, and a 3^0 pound weight with the object to be weighed, and a 3^6 pound weight in the other pan, since

$$716 = 3^6 - (3^2 + 3^1 + 3^0).$$

This fact could be written

$$716 = (1\ 000\ \bar{1}\bar{1}\bar{1})_3$$

with Barlow's notation of $\bar{1}$ for (-1). Barlow explained that every positive integer may be expressed in this modified base 3 notation (restricted to the three digits $\bar{1}$, 0, and 1) since the identity

$$2(3^k) = 3^{k+1} - 3^k$$

can be applied repeatedly to rewrite an ordinary base 3 string, such as $(222112)_3$, until every appearance of the digit 2 has been eliminated. This results in the modified base 3 notation, $(1\ 000\ \bar{1}\bar{1}\bar{1})_3$.

Barlow concluded his chapter with discussions of the relative merits of various bases, and the awkwardness of the impure decimal system used by the ancient Greeks. He considered base 12 to be the ideal base for all the usual reasons.

In 1814, Barlow's third publication on this topic appeared. This work consisted of certain entries ("Binary Arithmetic" and "Notation of the different Scales") in his *A New Mathe-*

matical and Philosophical Dictionary. These entries contained nothing significant on this topic beyond Barlow's earlier work.

Charles Hutton (1815)

Charles Hutton published his *A Philosophical and Mathematical Dictionary* in 1815. He explained under "Binary Arithmetic" that "this kind of Arithmetic was invented by Leibniz who pretended that it is better adapted than the common arithmetic, for discovering properties of numbers, and for constructing tables."

John Quincy Adams (1817)

In 1817 John Quincy Adams presented his *Report upon Weights and Measures* in compliance with a resolution of the Senate of March 3, 1817. This was a milestone on the path toward the 1866 legalization of the optional use of the decimalized international metric system within the United States.

John Leslie (1817)

The year 1817 also brought forth John Leslie's *The Philosophy of Arithmetic*, with a lengthy chapter entitled "Numeration." Leslie covered most of the then known results on nondecimal numeration for both whole and fractional numbers. To illustrate the flavor of his treatment the following is quoted:

> To reduce vulgar fractions to any scale, we have only to multiply the numerator by the root of that scale, and divide by the denominator; and to repeat this process, if requisite, on the successive remainders, till the quotients either terminate absolutely, or glide into circulation. Suppose it were sought to represent on the *Senary*, *Octary*, and *Denary Scales*, the fraction 355/113 or 3–16/113, which Peter Metius, a distinguished Dutch mathematician, and near relation of Adrian Metius of

Alkmaer, about the close of the sixteenth century, assigned for the approximate ratio of the circumference to the diameter of a circle.

After carrying out his announced plan, Leslie concluded that

$$355/113 = (3.0503301\ldots)_6$$
$$355/113 = (3.110376\ldots.)_8$$
$$355/113 = (3.141593\ldots.)_{10}$$

II. THE SECOND QUARTER

Heinrich Wilhelm Stein (1826)

In 1826 Stein published an article devoted entirely to a comparison of various numeration systems.[3] He argued that for base 10, one requires the concepts of 2, 3, 4, 5, 6, 7, 8, and 9 as well as 1, 10, 10^2, 10^3, and 10^4 in order to express all positive integers up to (but not including) 10^5—or a total of 13 concepts. In general, for base β, one would require $(x + \beta - 2)$ concepts to express numbers up to β^x. Letting $n = \beta^x$, the number of concepts could be expressed

$$\frac{\log n}{\log \beta} + \beta - 2 = f(\beta).$$

For a given n, that base would be best (require the least number of concepts) for which $f(\beta)$ would be a minimum. By Stein's criterion, if n is not too large, bases 2 and 3 fare poorly, 5 and 6 moderately well. Stein reported satisfaction with 10 on this basis and indicated no advantage in higher bases.

André-Marie Ampère (1838)

In 1838 appeared André-Marie Ampère's essay on the philosophy of science which included the author's classification scheme for the totality of human knowledge. Guiseppe

Peano was later to insist that this scheme represented a conscious application of Leibniz's binary system, but Ampère himself gave no such indication. Ampère partitioned all of human knowledge into two kingdoms and further divided each of these into two subkingdoms until there appeared eight cells in the third such partition. One of these eight is mathematics, whose further partitioning is shown in Figure 8.

Augustin-Louis Cauchy (1840)

Cauchy showed in 1840 (publication delayed until 1885) that decimal numeration may be modified analogously to the way Barlow had modified base 3 notation.[4] Consistent with Barlow's notation, Cauchy introduced $\bar{4}$ for (-4), $\bar{3}$ for (-3), and so on. He showed that the ten digits $\bar{4}$, $\bar{3}$, $\bar{2}$, $\bar{1}$, 0, 1, 2, 3, 4, and 5 could be used in place of the usual 10, since $6 = 1\bar{4}$, $7 = 1\bar{3}$, $8 = 1\bar{2}$, $9 = 1\bar{1}$. Cauchy argued that this modified decimal system would be simpler in some respects than the standard one. He called attention to the fact that $11^2 = 121$, $12^2 = 144$, and $13^2 = 169$ would be analogous (in appearance if not substance) to $1\bar{1}^2 = 1\bar{2}1$, $1\bar{2}^2 = 1\bar{4}4$, $1\bar{3}^2 = 1\bar{6}9$. He also showed that

$$1/7 = 0.142857142857\ldots$$
$$1/7 = 0.143\bar{1}\bar{4}\bar{3}143\bar{1}\bar{4}\bar{3}\ldots$$

where, in the latter, the second half of the period repeats the same digits as the first half, except for opposite sign. As further examples of this latter phenomenon, he cited

$$1/11 = 0.090909\ldots$$
$$1/11 = 0.1\bar{1}1\bar{1}1\bar{1}\ldots$$

and

$$1/13 = 0.076923076923\ldots$$
$$1/13 = 0.1\bar{2}\bar{3}\bar{1}231\bar{2}\bar{3}\bar{1}23\ldots$$

In conclusion Cauchy pointed out that in this modified base 10 notation, one could easily convert log N to log (1/N) by changing the sign of every digit.

- Arithmology
  - Elementary arithmology
    - Arithmography
    - Analysis
  - Megethology
    - Function theory
    - Probability theory
- Geometry
  - Elementary geometry
    - Synthetic geometry
    - Analytic geometry
  - Theory of forms
    - Theory of lines and surfaces
    - Molecular geometry

FIGURE 8. The Subdivisions of Mathematics According to Ampère's Classification Scheme of 1838.

D. Vicente Pujals (1844)

Outside of the mainstream of western mathematical scholarship, there appeared in 1844 *Filosofia de la Numeracion* by D. Vicente Pujals de la Bastida. This work extolled the virtues of base 12. Pujals traced the history of numeration and finally rejected, for one reason or another, all bases except 12. While most of the book is then devoted to pointing out the advantages of this base (simpler fractions, simpler divisibility rules, easier computation), it appears that Pujals would have favored base 12 regardless, since he considered this base God's choice. Are there not 12 lost tribes of Israel, 12 major prophets, 12 apostles, 12 major and minor scales of music to accompany the hymns to God? Pujals thought that for common use we should replace base 10 by base 12 as a prelude to a happier age living in consonance with God's plan. Scientists would have no difficulty making the change, but would the common man who is used to counting on his fingers? He could get used to counting the 12 joints of the four fingers of one hand.

III. THE THIRD QUARTER

Augustus DeMorgan (1853)

Augustus DeMorgan's *The Elements of Arithmetic* of 1853 is probably the earliest school text in the English language to include nondecimal numeration. DeMorgan wrote later that "The student should accustom himself to work questions in different systems of numeration, which will give him clearer insight into the nature of arithmetical processes than he could obtain by any other method."[5] In the 1853 text such different systems of numeration (bases 2, 5, and 12) are found not under the topic of "Numeration" but under "Scales of Notation." Under "Numeration" DeMorgan continued a long tradition of explaining base 10 numeration only.

Sir Isaac Pitman

In the February 9, 1856 issue of his *Phonetic Journal*, Sir Isaac Pitman printed an article entitled, "A New and Improved System of Numeration and Measurement," that is, an article extolling the virtues of the duodecimal system. As his brother later wrote:

> He seemed for years almost as hopeful of the adoption of the duodecimal scheme as of the success of the Writing and Spelling Reform; and of its ultimate general acceptance and use, he entertained no doubt. The three R's, reading, riting, and reckoning," he urged, would then become so easy and natural that their acquisition would indeed "come by nature."[6]

Aimé Mariage (1857)

With equal fervor Aimé Mariage came out in favor of base 8 in 1857.[7] Mariage's book essentially offered nothing beyond the arguments that Swedberg and his king had offered a century and a half earlier—except for more explicit examples of the advantages of this base in which 1/64 would be written 0.01 instead of the more awkward decimal fraction 0.015625.

John William Nystrom (1862)

In tune with Swedberg and Mariage, who wanted the base to be a power of 2, John William Nystrom, an American engineer, advocated base 16 in his 1862 publication entitled *Project of a New System of Arithmetic, Weight, Measure and Coins, Proposed to be Called the Tonal System with Sixteen to the Base*. He wrote on page 9:

> It is evident that 12 is a better number than 10 or 100 as a base, but it admits only one more binary division than 10, and would therefore, not come up to the general requirement.
>
> The number 16 admits binary division to an infinite extent, and would therefore, be the most suitable number as a base for arithmetic, weight, measure, and coins.

He wrote further, on page 11:

> The International Decimal Association is in favor of introducing the French metrical system, which is the most complete in existence, but has the evident disadvantages herein alluded to.

Nystrom advocated a 16 hour day, months of 16 or 17 days, a musical scale of 16 notes in the octave, a temperature scale with 256 (i.e. 16^2) degrees between the freezing and boiling points of water, and of course a hexadecimally divided circle.

DeMorgan, on page 371, asked how to explain why "an engineer who has surveyed mankind from Philadelphia to Rostof on the Don should for a moment entertain the idea of such a system being actually adopted, . . ."[8]

Four years later Nystrom came out in favor of the duodecimal system. His pamphlet *On the French Metric System: with a Discussion of a Duodecimal Notation* left no doubt that he wanted to prevent American adoption of the decimalized metric system. He objected to the length of the meter and its inability to be broken into four parts without involving fractional decimeters. At any rate, such a fourfolded meter would be inconvenient since it would not fit into an ordinary pocket. He also disliked the metric system because it was French; because it did not admit of binary division as required in practice; and because its use would necessitate altering drawings, patterns, taps, dies, reamers, mandrils, and so on.

E. Stahlberger (1869)

Without Barlow's convenient modified base 3 notation, Stahlberger analyzed the "either pan" weighing problem somewhat more deeply.[9] In his 1869 article, he specified the upper limit, $(3^{n-1} - 1)/2$ that may be weighed with the set of weights 1, 3, 3^2, ... , 3^n. He also provided a formula, namely

$$k = \frac{\log(2L + 1)}{\log(3)} - 1$$

for determining k, the highest power of 3 to be involved in weighing all integral weights up to L inclusive. For L = 1000, $k = 5.9^+$, indicating that $3^6 = 729$ should be the highest power of 3 involved.

A. Sonnenschein (1870)

An English language exercise book by Sonnenschein appeared in 1870 and included exercises in nondecimal numeration.[10] "Express 760 in each scale from the binary to the duodecimal scale" was one of them. Others required addition, multiplication, subtraction, and division to be performed in nondecimal scales. On page 75, under "Opinions of the Press," one finds the author identified as a "pupil, and a thoroughly taught pupil of Mr. DeMorgan."

J. W. L. Glaisher (1873)

In his article in the Messenger of Mathematics, Glaisher pointed out that the fact $1/81 = 0.\overline{012345679}\ldots$ can be generalized to all bases β. He proved:

Theorem 5.2: *For all bases* β, $1/(\beta - 1)^2 = 0.\overline{0123\ldots xy}\ldots$ where $x = \beta - 3$ *and* $y = \beta - 1$.

Glaisher's proof depends simply on the binomial expansion of $(\beta - 1)^{-2}$, which is

$$\beta^{-2} + 2\beta^{-3} + 3\beta^{-4} + \ldots + n\beta^{-n-1} + \ldots$$

whose coefficients are 1, 2, 3, ... , n, ... When this expansion is expressed in base β notation, the coefficients from $n = \beta$ on will require more than a single digit, resulting in a carrying pattern that produces the period 0123...xy. Here $x = \beta - 3$ and $y = \beta - 1$, so that the period runs through all base β digits except the digit $(\beta - 2)$.

Hermann Hankel (1874)

Hankel worked out further details on Stein's criterion for "best base," and provided the following table, showing the number of concepts required to express all whole numbers up to M inclusive for selective bases β:[11]

For β =	2	3	4	5	6	10	12	20
For M = 10^3	11	8	8	8	9	12	13	21
For M = 10^6	21	15	12	13	13	15	17	24

Thus, for M = 10^6, base 4 comes out best.

IV. THE FOURTH QUARTER

Moritz Cantor (1875)

Moritz Cantor reported that carnival booths were selling a "Tell Your Age" game for 12 Pfennig and that the game astonished him until he discovered it to be an application of binary numeration.[12] The game consisted of seven cards having the numbers from 1 to 100 distributed over them—some numbers appearing on more than one card. Card I had the key number 1 associated with itself, II had 2, III had 4, IV had 8, V had 16, and VII had 64. A 37-year-old person would see his age listed on cards I, III, and VI and so inform the holder of the cards. The cardholder would simply add the key numbers 1, 4, and 32 to arrive, astonishingly, at 37. As Cantor analyzed the game, every number up to 100 could be represented by a 7-bit string, 37 for example by 0100101. The number 37 would therefore appear on cards I, III, and VI in accordance with which bits (starting from the right end of the string) were "1"s rather than "O"s. Of course, the seven cards would have sufficed to extend the range of possible ages to 127, the highest number that can be expressed as a 7-bit string, namely 1 111 111.

Felix Müller (1876)

Dr. Felix Müller replied to Cantor within the year in the same journal.[13] Müller contended that as early as 1859, the 11th graders of a certain secondary school in Berlin had dealt with this problem as a by-product of proving the identity:

$$\frac{1-x^2}{1-x} \cdot \frac{1-x^4}{1-x^2} \cdot \frac{1-x^8}{1-x^4} \cdot \frac{1-x^{16}}{1-x^8} \cdot \frac{1-x^{32}}{1-x^{16}} = \frac{1-x^{32}}{1-x} \tag{5.1}$$

which in turn implies the identity:

$$(1+x)(1+x^2)(1+x^4)(1+x^8)(1+x^{16}) = 1+x+x^2+x^3+x^4+x^5+ \ldots +x^{31}. \tag{5.2}$$

It should be noted, argued Müller, that the exponents of the right side of 5.2 include every whole number from 1 to 31. Since these must be the result of carrying out the indicated multiplication of the left side of 5.2, it follows that each exponent on the right is the sum of some combination of exponents of the left. Hence every whole number from 1 to 31 inclusive can be expressed as the sum of some combination of the first 5 powers of 2, namely 1, 2, 4, 8, and 16.

Müller gave as one possible interpretation of this last fact the explanation that a set of 5 weights (consisting of 1, 2, 4, 8, and 16) would suffice to weigh all objects of integral weights up to 31. By dealing with generalized versions of 5.1 and 5.2, Müller indicated that the results could readily be extended up to $2^n - 1$ for any n and certainly to $2^7 - 1 = 127$, as required by those carnival cards.

Moreover, Müller reported, those 11th graders had designed a set of 5 cards containing the numbers from 1 to 121 inclusive, based on powers of 3, and sometimes taken with a minus sign. However, the reader of the cards would have to specify not only which of the five cards contained his age (or other number he was temporarily keeping secret), but whether the number appeared in light or bold print—the latter indicating that a minus sign would have to be used. Thus, 49, for example, appears light on I and II, bold on III and IV, and light on V, since

$$49 = 3^0 + 3^1 - 3^2 - 3^3 + 3^4,$$

a fact which Barlow would have written

$$40 = (1\bar{1}\bar{1}11)_3.$$

This latter game, Müller indicated, was a by-product of dealing with the identity

$$(5.3)\qquad \frac{1-x^3}{1-x}\cdot\frac{1-x^9}{1-x^3}\cdot\frac{1-x^{27}}{1-x^9}\cdot\ \dots\ \cdot\frac{1-x^{3^{n+1}}}{1-x^{3^n}} = \frac{1-x^{3^{n+1}}}{1-x}$$

which implies the identity

$$(5.4)\qquad \begin{aligned}&(1+x+x^2)(1+x^3+x^6)(1+x^9+x^{18})\ \dots\ (1+x^{3^n}+x^{2\cdot 3^n})\\ &\qquad = 1+x+x^2+x^3+x^4+\ \dots\ +x^{3n}\ .\end{aligned}$$

If 5.4 is now divided by x^k, where $k = (1/2)(3^{n+1} - 1)$ then the identity

$$(5.5)\qquad \begin{aligned}&(x^{-1}+1+x^1)(x^{-3}+1+x^3)(x^{-9}+1+x^9)\dots(x^{-3^n}+1+x^{3^n})\\ &\qquad =x^{-k}+\dots+x^{-2}+x^{-1}+1+x^1+x^2+x^3+\dots+x^k\end{aligned}$$

results. For $n = 4$, for example, 5.5 gives the desired result that all numbers from 1 to 121 can be expressed in terms of 3^0, 3^1, 3^2, 3^3, and 3^4—the coefficients -1, 0, and $+1$ being permitted.

Müller made no reference to Barlow, who had assumed the possibility of standard base 3 notation for all whole numbers and then had shown that each base 3 string could be modified into his modified base 3 string. Müller managed to bypass that assumption and prove the desired result directly. The five Müller cards appear in Figure 9.

Cantor might have been further humbled if it had been pointed out to him that the 1814 edition of Jacques Ozanam's *Recreations in Mathematics and Natural Philosophy* already contained a description of a set of "Guess the number cards," with the explanation that the trick involved the numbers 1, 2, 4, 8, etc.

P. A. MacMahon (1886, 1891)

Starting with a series of algebraic identities that included the ones Müller had used, MacMahon did a more complete analysis of the two problems:[14]

(1) To assign a series of weights so as to be able to weigh any weight of an integral number of pounds from 1 to n in-

I			
1	32	62	92
2	*34*	*64*	*94*
4	35	65	95
5	*37*	*67*	*97*
7	38	68	98
8	*40*	*70*	*100*
10	41	71	101
11	*43*	*73*	*103*
13	44	74	104
14	*46*	*76*	*106*
16	47	77	107
17	*49*	*79*	*109*
19	50	80	110
20	*52*	*82*	*112*
22	53	83	113
23	*55*	*85*	*115*
25	56	86	116
26	*58*	*88*	*118*
28	59	89	119
29	*61*	*91*	*121*
31			

II			
2	32	61	*93*
3	33	*65*	*94*
4	34	*66*	95
5	*38*	*67*	96
6	*39*	68	97
7	*40*	69	*101*
11	41	70	*102*
12	42	*74*	*103*
13	43	*75*	104
14	*47*	*76*	105
15	*48*	77	106
16	*49*	78	*110*
20	50	79	*111*
21	51	*83*	*112*
22	52	*84*	113
23	*56*	*85*	114
24	*57*	86	115
25	*58*	87	*119*
29	59	88	*120*
30	60	*92*	*121*
31			

III			
5	*35*	*64*	*93*
6	*36*	*65*	*94*
7	*37*	*66*	95
8	*38*	*67*	96
9	*39*	68	97
10	*40*	69	98
11	41	70	99
12	42	71	100
13	43	72	101
14	44	73	102
15	45	74	103
16	46	75	*113*
17	47	76	*114*
18	48	*86*	*115*
19	49	*87*	*116*
20	*59*	*88*	*117*
21	*60*	*89*	*118*
22	*61*	*90*	*119*
32	*62*	*91*	*120*
33	*63*	*92*	*121*
34			

IV			
14	*35*	55	*102*
15	*36*	56	*103*
16	*37*	57	*104*
17	*38*	58	*105*
18	*39*	59	*106*
19	*40*	60	*107*
20	41	61	*108*
21	42	62	*109*
22	43	63	*110*
23	44	64	*111*
24	45	65	*112*
25	46	66	*113*
26	47	67	*114*
27	48	*95*	*115*
28	49	*96*	*116*
29	50	*97*	*117*
30	51	*98*	*118*
31	52	*99*	*119*
32	53	*100*	*120*
33	54	*101*	*121*
34			

V			
41	*62*	*82*	*102*
42	*63*	*83*	*103*
43	*64*	*84*	*104*
44	*65*	*85*	*105*
45	*66*	*86*	*106*
46	*67*	*87*	*107*
47	*68*	*88*	*108*
48	*69*	*89*	*109*
49	*70*	*90*	*110*
50	*71*	*91*	*111*
51	*72*	*92*	*112*
52	*73*	*93*	*113*
53	*74*	*94*	*114*
54	*75*	*95*	*115*
55	*76*	*96*	*116*
56	*77*	*97*	*117*
57	*78*	*98*	*118*
58	*79*	*99*	*119*
59	*80*	*100*	*120*
60	*81*	*101*	*121*
61			

FIGURE 9. The Five Müller Cards of 1876.

clusive, the weights being placed in only one scale-pan; and

(2) The same problem when the weights may be placed in *either* of the two scale-pans.

MacMahon further restricted the problems by insisting that no other weighings except those from 1 to n should be possible and that each weighing is to be possible in only one way. Barlow and Müller had also tacitly assumed these restrictions.

MacMahon's first paper appeared in 1886, but seemed to have made no impression on the editor of London's prestigious science journal *Nature*, which published a long series of "weighing problem" letters in 1890–91. These letters, however, barely brought the *Nature* reader up to the 1811 Barlow level. MacMahon's *Nature* article put an end to the series of letters. He showed for N = 40, for example, that the second weighing problem had the following eight solutions:

1^{40}	40 weights
$1, 3^{13}$	14 weights
$1^4, 3^{13}$	8 weights
$1, 3, 9^4$	6 weights
$1^{13}, 27$	14 weights
$1, 3^4, 27$	6 weights
$1^4, 9, 27$	6 weights
$1, 3, 9, 27$	4 weights

The "exponents" indicate how many of each weight are represented.

The last of these eight solutions corresponds to the Barlow-Müller solution and is hereby shown to be the best solution, that is, it requires the least number of weights. Among MacMahon's results (announced in each of the three papers) are:

Theorem 5.3: *Weighing problem* (*1*) *has* 2^{s-1} *solutions if* $n + 1 = p^s$, *where* p *is prime.*

Theorem 5.4: *Weighing problem* (2) *has* 2^{s-1} *solutions if* $2n + 1 = p^s$, *where* p *is prime.*

MacMahon showed that the former of the two theorems is connected with the problem of how many ways the expression

$$1 + x + x^2 + x^3 + x^4 + x^5 + \ldots + x^n$$

may be factored. Similarly, the latter theorem is connected with the question of how many ways

$$x^{-n} + \ldots + x^{-2} + x^{-1} + 1 + x^{2} + x^{3} + x^{4} + \ldots + x^{n}$$

may be factored.

According to Leopold Gegenbauer, the algebraic identities which formed the bases of Müller's and MacMahon's work were already used by Euler (p. 275, *Introductio in analysin infinitorum*) in showing that every whole number may be represented uniquely as a polynomial in 3 using only 3 coefficients, −1, 0, and 1.

E. Collignon (1897)

The Barlow and Cauchy modifications of base β notation involved negative digits, but for each negative digit introduced, a positive digit was deleted leaving the total number of digits equal to β. Such modification is not possible for base 2, since for the introduction of $\bar{1}$, one cannot dispense with the only positive digit 1. Collignon investigated the use of base 2 modified to permit the *three* digits, $\bar{1}$, 0, and 1.[15]

Collignon found that such modified base 2 representations were not unique, since, for example,

$$(360)_{10} = (101101000)_2 = (110\bar{1}01000)_2 = (10\bar{1}0\bar{1}01000)_2$$

and since

$$\begin{aligned}(15827)_{10} &= (11110111010011)_2 = (10000\bar{1}100\bar{1}01110\bar{1})_2 \\ &= (10000\bar{1}00\bar{1}01010\bar{1})_2.\end{aligned}$$

After several additional examples, he announced his conviction that it was always possible to find one modified base 2 string in which the digit 1 (taken positively or negatively) would be separated from the next ±1 by one or more 0's. He indicated that this was also true of rational fractions, since

$$\begin{aligned}1/17 &= 0.000011110000111100001111\ldots \\ &= 0.0001000\bar{1}0001000\bar{1}0001000\bar{1}\ldots\end{aligned}$$

and since

$$\begin{aligned} 4/9 &= 0.011100011100011000\ldots \\ &= 0.100\bar{1}00100\bar{1}00100\bar{1}001\ldots \end{aligned}$$

Collignon maintained that the usual four fundamental operations could be performed more expeditiously in his modified base 2 notation than in the standard base 2, since fewer nonzero digits would be involved.

He also noted that

$$\begin{aligned} 2^{3k} &\equiv 1 \pmod 7 \\ 2^{3k+1} &\equiv 2 \pmod 7 \\ 2^{3k+2} &\equiv 4 \pmod 7, \end{aligned}$$

and hence that one could readily determine divisibility by 7 for any N represented as a binary string (standard or modified). For $N = (100111010011)_2 = (10100\bar{1}01010\bar{1})_2$, for example, the standard string indicates that

$$N = 2^{11} + 2^8 + 2^7 + 2^6 + 2^4 + 2^1 + 2^0,$$

and hence that

$$\begin{aligned} N &\equiv 4 + 4 + 2 + 1 + 2 + 2 + 1 \\ &\equiv 2 \pmod 7. \end{aligned}$$

Similarly, the modified string indicates that

$$N = 2^{11} + 2^9 - 2^6 + 2^4 + 2^2 - 2^0$$

and hence that

$$\begin{aligned} N &\equiv 4 + 1 - 1 + 2 + 4 - 1 \\ &\equiv 2 \pmod 7. \end{aligned}$$

VITTORIO GRÜNWALD (1885)

Grünwald investigated the use of $\beta = -10$ with the usual ten digits, 0, 1, ... , 8, and 9.[16] In this nonstandard system,

$$\begin{aligned} (15937)_{-10} &= 7 + 3(-10) + 9(-10)^2 + 5(-10)^3 + (-10)^4 \\ &= (5877)_{10}. \end{aligned}$$

A	B	C	D
0	zero	| |	0000
1	one	| !	0001
2	two	| ¡	0010
3	three	| :	0011
4	four	! |	0100
5	five	! !	0101
6	six	! ¡	0110
7	seven	! :	0111
8	eight	¡ |	1000
9	nine	¡ !	1001
10	ten	¡ ¡	1010
11	onze	¡ :	1011
12	douze	: |	1100
13	treize	: !	1101
14	quorze	: ¡	1110
15	quinze	: :	1111

FIGURE 10. Benjamin Peirce's 'Improved' Binary Notation (column C) Compared with Leibniz's (column D). Column B shows Peirce's proposed nomenclature.

He patiently investigated the four fundamental operations as well as extraction of square and cube roots in this system. He discussed at length how to find (−N) given (+N) and presented numerous humble results such as $(N)^3 + (-N)^3 = (N)^3 - (N)^3 = 0$ and that division by $(-10)^3$ would have the effect of moving the fraction marker three places toward the left.

A series of miscellaneous publications appearing during the last quarter of the nineteenth century will now be discussed in chronological order.

Benjamin Peirce (1876)

The 1876 *United States Coast Survey Report* included as Appendix 6, Benjamin Peirce's "A New System of Binary Arithmetic." Peirce considered his system of dots and vertical dashes more economical of space and a distinct improvement over Leibniz's version of the binary system. He chided Leibniz for not having provided any system of nomenclature and proceeded to remedy this defect. Figure 10 shows Peirce's notation and nomenclature for the first 16 whole numbers. Additional nomenclature included special names for each quadrate (group of 4 binary digits), namely units, ties, tries, quads, quints, sies, septs, and octs. His grouping into fours suggests ready interpretation as binary-coded base 16.

Benjamin Peirce was a highly regarded American mathematician who held a professorship at Harvard. His son, Charles, also worked with binary expressions, which he called "secundals." Charles' work was only in manuscript form until 1976, when Carolyn Eisele published it.

G. Bellavitis (1877)

Bellavitis reduced long binary strings by indicating the numbers of "0"s and "1"s that appeared in succession.[17] Thus, the repeating binary fraction $1/13 = 0.\overline{0001\ 0011\ 1011}\ldots$ was indicated by 312312, which was further shortened in his final table to 312, since the second half of the period could be predicted from the first. According to Dickson, "G. Bellavitis noted that the use of base 2 renders much more compact and convenient Gauss's table and hence constructed such a table."[18] The Gauss table in question is Table 3 of *Disquisitiones Arithmeticae* (a more extended version was published posthumously), which gives the decimal periods of fractions of the type 1/p, where p is prime.

LLOYD TANNER (1878)

Tanner reported on numbers N whose n β-adic digits form the tail end of the β-adic string of N^2, that is, numbers N which satisfy

$$N^2 - N = K\beta^n$$

for some positive integer K.[19] He determined that for $\beta = 6$, 10, and 12 there exist only two such N each, namely

N = 3221350213	and N = 3334205344	for base 6
N = 8212890625	and N = 1787109376	for base 10
N = 21B61B3854	and N = 9A05A08369	for base 12

where A and B represent 9 + 1 and 9 + 2 respectively.

FRANZ HOCEVAR

According to Dickson, Hocevar noted that N = 104533 or $N = (11001100001010101)_2$ is divisible by 17 since 0101 + 1000 − 1001 + 1 = 0, or more generally, that if N is written to the base β and then separated into groups (starting from the right) G_0, G_1, G_2, ... each of q digits, then N is divisible by $\gamma + 1$ (where $\gamma = \beta^q$) if $G_0 + G_1 + G_2 + \dots$ is divisible by $\gamma + 1$.[20] This is not startling if one notes that the grouping has the effect of rewriting N to the base γ, and that the G_i's are the base γ digits of N. The insight involved was already displayed by Legendre, who had interpreted a base 2 string as a "base 2 coded base 64" string (see Theorem 4.7). After such interpretation the usual divisibility rule (for divisibility by $\gamma + 1$) applies.

Brockhaus (1883)

The 1883 or 13th edition of *Brockhaus Conversations-Lexicon* carried under "Dyadik" a brief description of the binary system. This is not noteworthy in itself, since many earlier encyclopaedias in English, French, and German had done the same. *Brockhaus* seems, however, to be the first publication of any type to take notice of the fact that Bishop Juan Caramuel had published on the binary system some 30 years before Leibniz.

Charles Berdellé (1887)

Charles Berdellé proposed the eight symbols shown below as the eight octal digits:[21]

decimal :	0	1	2	3	4	5	6	7
binary :	000	001	010	011	100	101	110	111
Berdellé octal :	 ─	┐	┬	┬┐	┌	┌┐	┌┬	┌┬┐

Obviously, Berdellé's octal digits are but thinly disguised 3-bit binary strings. Berdellé declared that it was easy to make the two systems (binary and octal) one and the same. This amounts simply to the Legendre-Hocevar insight that a binary string is at once a binary-coded octal string. Berdellé also proposed a system of nomenclature for his octal system.

Oskar Simony (1887)

Simony reported that he found binary notation more suitable than decimal—at least for certain purposes—in his topological investigation of knots.[22] In this connection Simony reviewed Caramuel's and Leibniz's work on the binary system, including the binary interpretations of the Figures of Fohy.

Edouard Lucas (1891)

According to Dickson, Edouard Lucas used binary notation to prove $2^{31} - 1$ prime.[23] His 1891 *Théorie des nombres* continued a tradition started by Barlow's 1811 book giving numeration systems a definite place in books on the theory of numbers. Lucas's 1891 *Récréations mathématique* has already been cited for a probably false reference to Simon Stevin as a duodecimal advocate. On the topic of numeration systems, the two Lucas books together offered little beyond Barlow's 1811 book.

Hermann Scheffler (1891)

Scheffler made some contributions to the factorization of numbers of the type $2^n + 1$ by writing possible factors in binary notation.

Georg Cantor (1895)

In 1895 Georg Cantor (the set theorist, not Moritz Cantor the mathematical historian) published a proof of

$$c = 2^{\aleph_o}$$

based on the fact that c is the cardinal number of the continuum as well as the interval (0,1), and that every number x in that interval can be expressed in the form

$$x = \frac{f(1)}{2^1} + \frac{f(2)}{2^2} + \frac{f(3)}{2^3} + \cdots + \frac{f(v)}{2^v}$$

where f(v) is either 0 or 1. The cardinal number associated with the set of positive integers is $\aleph_o$. The form of x shown is, of course, expanded binary notation.

F. J. Studnicka (1896)

Studnicka reported that the property displayed by

$$\begin{array}{r} 9\,8\,7\,6\,5\,4\,3\,2\,1 \\ -\,1\,2\,3\,4\,5\,6\,7\,8\,9 \\ \hline 8\,6\,4\,1\,9\,7\,5\,3\,2, \end{array}$$

where every nonzero decimal digit is represented in each of the three lines in a predictable order, can be generalized to other β-adic systems, where β is even.[24]

Studnicka gave as examples:

$$\begin{array}{r} 3\,2\,1 \\ -\,1\,2\,3 \\ \hline 1\,3\,2 \end{array} \qquad \text{for base 4}$$

$$\begin{array}{r} 5\,4\,3\,2\,1 \\ -\,1\,2\,3\,4\,5 \\ \hline 4\,1\,5\,3\,2 \end{array} \qquad \text{for base 6}$$

$$\begin{array}{r} 7\,6\,5\,4\,3\,2\,1 \\ -\,1\,2\,3\,4\,5\,6\,7 \\ \hline 6\,4\,1\,7\,5\,3\,2 \end{array} \qquad \text{for base 8}$$

$$\begin{array}{r} \mathrm{B}\,\mathrm{A}\,9\,8\,7\,6\,5\,4\,3\,2\,1 \\ -\,1\,2\,3\,4\,5\,6\,7\,8\,9\,\mathrm{A}\,\mathrm{B} \\ \hline \mathrm{A}\,8\,6\,4\,1\,\mathrm{B}\,9\,7\,5\,3\,2 \end{array} \qquad \text{for base 12}$$

He believed that through open display and proof of the property in question, the opportunity offered itself to acquaint students with different numeration systems.

Robert M. Pierce (1898)

Pierce published an essay in which he proposed that mathematicians and others be polled to ascertain the best possible base; serious steps should then be taken to replace base 10 by that best one.[25]

Giuseppe Peano (1899)

In 1899 Giuseppe Peano (the *very* Peano after whom the Peano Axioms are named) published an article proposing a new system of stenography that had the binary system as its foundation. Apparently he preferred "." and ":" for "0" and "1" respectively and showed the first few positive integers as follows:

:   :.   ::   :..   :.:   ::.   :::   :...

This article has been cited earlier in connection with Peano's insistence that Lamarck's and Ampère's classification schemes represented applications of the binary system. Peano wrote:

> A keyboard machine that writes a syllable at a time is that of Michela and is in use in our Senate. It uses the [1023 possible] combinations of 10 keys, associating a numerical value with each combination. A technical commission appointed by the Senate (January 30, 1880) and the Chamber of Deputies declared that with this machine "they were able to handle satisfactorily the ordinary stenographic transcriptions with respect to both speed and accuracy."
>
> Now the [proposed] binary writing is notably simpler and faster than that of Michela's machine. Using both hands, . . . , one can write with a single stroke 16 binary symbols, or two syllables—forming 65,536 combinations.
>
> The above-mentioned binary writing needs the knowledge of a special alphabet, which is not more difficult to learn than the alphabet of any people, or the stenographic alphabet, or the location of a drawer in a stenographic box. It is especially simple to learn since the symbols are constructed on the basis of a general principle.
>
> It has all the advantages for reading and ordinary writing. Using a small, quite simple machine, one can write with a speed which is superior to the speed of ordinary stenography. It can be sent by wire, utilizing all the potential of the telegraphic wire, things that the machines of Baudot and Ostrogowich do not do completely yet. And if some of the preceding applications will be of common use, I shall show how it can be printed with ordinary printing presses. These and other advantages derive from the pure application of binary numeration.[26]

Peano introduced a symbol that may be described as an octagonal star or an 8-petal daisy. Each petal present represents "1", or when absent "0", thus setting up a one-to-one correspondence between the 2^8 8-bit strings and the 2^8 daisies. Thus, for example,

11111111 =

01111111 =

11111110 =

00000101 =

10101010 =

00000001 =

The petals on each daisy are numbered counterclockwise starting from the southeast petal (whose place value is 2^0) to the south petal (2^7).

The daisies would be typed two daisies at a time with Peano's proposed stenographic machine. Since most typists have fewer than 16 fingers, presumably the keys are so arranged that one finger could strike more than one key simultaneously. While each daisy was to represent an *Italian* syllable, Peano thought that appropriate assignments could readily be made in other languages also. Twenty-five of the 2^8 or 256 daisies were also assigned letters of the alphabet and ten of them were to represent the ten decimal digits. Peano's list of these 35 assignments showed only letters and digits together with their associated daisies. In Figure 11, however, the daisies are omitted in favor of their binary, hexadecimal, and decimal equivalents.

Before Peano introduced his daisies in this article, he pointed out that the International Morse Code (which assigns 5 symbols to each decimal digit) would require 14,445 symbols to transmit all numbers from 1 to 999. However the binary system of 0s and 1s could do the same job with only 8977 symbols. These figures seem somewhat irrelevant since he then assigned an 8-bit daisy to each decimal digit.

A minor difficulty with Peano's daisies is that one cannot

Binary	Hexadecimal	Decimal	Meaning Assigned by Peano
0000 1000	08	008	A or 8
0100 0000	40	064	B
1110 0000	E0	224	C
0010 0000	20	032	D
0001 1000	18	024	E
1100 0100	C4	196	F
0110 0000	60	096	G
1000 0000	80	128	H
0001 0000	10	016	I
1001 0000	90	144	J
1100 1000	C8	200	K
0000 0010	02	002	L or 2
0000 0001	01	001	M or 1
1000 0001	81	129	N
0000 1100	0C	012	O
1100 0000	C0	192	P
1110 0100	E4	228	Q
1000 0010	82	130	R
0000 0011	03	003	S or 3
1010 0000	A0	160	T
0000 0100	04	004	U or 4
0100 0100	44	068	V
0110 0011	63	099	X
0001 0100	14	020	Y
0010 0011	23	035	Z
0000 0001	01	001	1
0000 0010	02	002	2
0000 0011	03	003	3
0000 0100	04	004	4
0000 0101	05	005	5
0000 0110	06	006	6
0000 0111	07	007	7
0000 1000	08	008	8
0000 1001	09	009	9
0000 1010	0A	010	0 or 10

FIGURE 11. Numeric Values (in Binary, Hexadecimal, and Decimal) Assigned to 25 Letters of the Alphabet and the Ten Decimal Digits by Peano.

distinguish the (0000 0000) daisy from the complete absence of a daisy. While an obvious pattern emerges in the assignment of daisies to decimal digits, no readily discernible pattern appears among the assignments for the 25 letters. No justification is apparent as to why five of the letters should have to share their daisies with digits.

A second publication of Peano's dealing with this topic is his monumental *Formulaire de mathématiques* (in French, published in 1901). Pages 75–78 of Tome 3 are devoted to both ancient and modern systems of numeration. Here the daisies are again explained and the example

$$1900 = (0000\ 0111)\ (01101100)$$

is given to show that two daisies would suffice to represent 1900. By his earlier scheme, displayed in Figure 11, 1900 would have required 4 daisies, i.e.,

$$1900 = (0000\ 0001)(0000\ 1001)(0000\ 1010)(0000\ 1010).$$

In repeating some of the assignments of binary symbols to letters of the alphabet, Peano now dispensed with the daisies, and indicated the assignments as shown by the following examples:

.... !... = A
!... .... = F
....! = M
!!!. .... = K

Fourteen such assignments are given, all of them consistent with those indicated in Figure 11 except F and K. His use of "!" for "1" is also new.

While giving the histories of the numbers e and π (on pages 154 and 177 respectively) Peano indicated that

$$e = (10.10110111\ 11100001\ 01010001\ \ldots)_2 \text{ and}$$
$$\pi = (11.00100100\ 00111111\ 01101010\ \ldots)_2.$$

Alfred B. Taylor (1887)

Discussions on "which base is best" continued throughout the last quarter of the nineteenth century. The most comprehensive of these was a 70-page paper read by Alfred B. Taylor before the American Philosophical Society on October 21, 1887.[27] The paper was in part a plea to reject the decimalized metric system considered by Taylor to be beautiful, but of no use to the tradesman or businessman. He raised the question whether the well-nigh universal decimal numeration was really natural. He decided it was natural in the same sense that ignorance was. He came to the conclusion that the "octonary" (base 8) scale would be the best possible one.

William Woolsey Johnson (1891)

Johnson came to the same conclusion in his article published in the immediate predecessor of the *Bulletin of the American Mathematical Society*.[28] He stated:

> Now it is to be noticed that if the radix is a power of two, we have virtually all the advantages of the binary system. For example, if we have a number expressed in the octonary system, we have only to substitute for the characters 0, 1, 2, ... , 7 their binary equivalents 000, 001, 010, ... , 111 to obtain the number in the binary system.
>
> The digital expression of a number in the octonary system would be much more suggestive of its intrinsic nature than expression in any non-binary system, for the highest power of two contained as a factor in a number is its most important characteristic. Again the distinction between numbers of the form 4n + 1 and those of the form 4n + 3 is of great importance in the theory of numbers, and in the octonary system it would be obvious at a glance to which of these classes a given uneven number belongs. So also with the distinction between "evenly even" and "unevenly even" numbers. It is interesting also to note that the square of every uneven number would end in 1, the preceding figures expressing a triangular number. Thus the uneven squares in octonary notation are 1, 11, 31, 61, 101, . . . As there is no doubt that our ancestors originated the

decimal system by counting on their fingers, we must, in view of the merits of the octonary system, feel profound regret that they should have perversely counted their thumbs, although nature had differentiated them from the fingers sufficiently, she might have thought, to save the race from this error.[28]

E. GELIN (1896)

Gelin also considered base 8 the best.[29] He favored 8 over 10 and 12, for example, since the successive powers of 8 contain more divisors (proportional to their value), than do the same powers of 10 or 12. He pointed out that 8^n, 10^n, and 12^n, have, respectively,

$$3n + 1, \quad n^2 + 2n + 1, \quad 2n^2 + 3n + 1$$

divisors and that the following inequalities hold:

$$\frac{3n + 1}{n^2 + 2n + 1} > \frac{8^n}{10^n}$$

$$\frac{3n + 1}{2n^2 + 3n + 1} > \frac{8^n}{12^n}$$

T. N. THIELE (1889)

Thiele favored base 4.[30] He gave the usual reasons for wanting a base that is a power of 2. His choice of 4 over other powers of two rested on pedagogical considerations. First he analyzed the number of "results" pupils would need to memorize for various bases. For base 4, for example, out of the usual 16 multiplication facts, he considered only the following 3 as nontrivial: $2 \cdot 2 = 10$, $2 \cdot 3 = 12$, and $3 \cdot 3 = 21$. Base 2, by Thiele's reasoning, has no nontrivial multiplication facts at all. In general, base β has, respectively,

$$\frac{\beta(\beta - 1)}{2} \quad \text{and} \quad \frac{(\beta - 1)(\beta - 2)}{2}$$

nontrivial addition and multiplication facts. By these formulae, bases 2, 4, 6, 10 and 16 involve, respectively, 1, 9, 25, 81, and 225 "results" (nontrivial addition and multiplication facts). By replacing base 10 by 4 for common use, even the dullest students could memorize the 9 results involved. The brighter ones could extend the addition and multiplication tables to, say $13 + 13 = 32$ and $13 \cdot 13 = 301$ (tantamount to working in base 8). Still brighter students could make the extensions tantamount to working in base 16 or even 32. Thus argued Thiele every student could pick that base best suited to his or her talents and needs.

Herbert Spencer (1896)

The English philosopher Herbert Spencer came out in favor of base 12 for common use in a passionate plea against British adoption of the decimalized metric system. The gist of his argument echoes through his last paragraph:

> See, then, the strange position. The vast majority of our population consists of working people, people of narrow incomes and the minor shopkeepers who minister to their wants. And these wants daily lead to myriads of purchases of small quantities for small sums, involving fractional divisions of measures and money—measuring transactions probably fifty times as numerous as those of the men of science and the wholesale traders put together. These two small classes, however, unfamiliar with retail measuring transactions, have decided that they will be better carried on by the metric system than by the existing system. Those who have no experimental knowledge of the matter propose to regulate those who have! The methods followed by the experienced are to be rearranged by the inexperienced![31]

Edward Brooks (1876)

Brooks devoted pages 113–134 of his *The Philosophy of Arithmetic* to a discussion of numeration systems. He made a strong plea for replacing the decimal system by the duodecimal for common use.

Neither the pleas of those who wanted the decimal system replaced for common use, nor of those who wanted schools to take up the topic of nondecimal numeration regardless of which base was in common use, seemed to have had any appreciable effect on the content of the school texts of the last quarter of the nineteenth century.

School Texts (1875–1899)

Of course, the later editions of the Sonnenschein arithmetic previously mentioned continued to carry some exercises dealing with nondecimal numeration, as might be expected of the author, who has been identified as a pupil of DeMorgan. The *Higher Algebra* of Hall and Knight has already been mentioned in a quotation by W. W. Sawyer as an English text that covered this topic. The great majority of school texts in arithmetic and algebra restricted their discussion of "numeration" to decimal numeration. From the Müller report, it may well be assumed that some individual teachers included nondecimal numeration, at least as a by-product of certain other discussions in mathematics.

NOTES TO CHAPTER V

[1]Carl Friedrich Gauss, *Disquistiones arithmeticae*, trans. Arthur A. Clarke, (New Haven: Yale University Press, 1966), p. 377.

[2]Dietrich Mahnke, "Leibniz auf der Suche nach einer allgemeinen Primzahlgleichung," *Bibliotheca mathematica* (3)13 (1912–13):48.

[3]Heinrich Wilhelm Stein, "Über die Vergleichung der verschiedenen Numerationssysteme," *Journal für die reine und angewandte Mathematik* 1 (1826):369–371.

[4]Augustin-Louis Cauchy, *Oeuvres*, (Paris: Gauthier-Villars et Fils, 1885), 5:431–442.

[5]Augustus DeMorgan, *Study and Difficulties of Mathematics*, (Chicago: Open Court Publishing Company, 1910), p. 7.

[6]Benn Pitman, *Sir Isaac Pitman*, (Cincinnati: C. J. Krehbiel Company, 1902), p. 189.

[7]Aimé Mariage, *Numération par huit*, (Paris: Le Normant, 1857).

[8]Augustus DeMorgan, *A Budget of Paradoxes*, (London: Longmans, Green, and Company, 1872).

[9]E. Stahlberger, "Über einen Gewichtssatz, dessen Gewichte nach Potenzen von 3 geordnet sind," *Repertorium für Experimental Physik, für Physikalische Technik, Mathematische und Astronomische Instrumentenkunde* 5 (1869):10–13.

[10]A. Sonnenschein and H. A. Nesbitt, *The Science and Art of Arithmetic, Exercise Book, Part I*, (London: W. Swan Sonnenschein and Allen, c. 1870).

[11]Hermann Hankel, *Zur Geschichte der Mathematik in Alterthum und Mittelalter*, (Leipzig: B. G. Teubner, 1874).

[12]Moritz Cantor, "Zahlentheoretische Spielereien," *Zeitschrift für Mathematik und Physik* 20 (1875):134–135.

[13]Felix Müller, "Über eine Zahlentheoretische Spielerei," *Zeitschrift für Mathematik und Physik* 21 (1876):227–28.

[14]P. A. MacMahon, "Certain Special Partitions of Numbers," *Quarterly Journal of Pure and Applied Mathematics* 21 (1886):367–373; and "The Theory of Perfect Partitions of Numbers and the Compositions of Multipartite Numbers," *Messenger of Mathematics* 20 (1891):101–119; and "Weighting by a Series of Weights," *Nature* 43 (1891):113–14.

[15]E. Collignon, "Note sur l'arithmétique binaire," *Journal de mathématiques elémentaires* 21 (1897):101–106, 126–131, 148–151, 171–174.

[16]Vittorio Grünwald, "Intorno all'arithmetica dei sisteme numerici a base negativa," *Giorgnale di matmatiche di Battaglini* 23 (1885):203–221, 367.

[17]G. Bellavitis, "Sulla risoluzione delle congruenze numeriche e sulle tabole che danno i logaritme (indici) degli interi rispetto ai vari moduli," *Atti, Accademia nationale dei Lincei, Roma, classe di scienze fisiche, matmatiche e naturali*, series 3, 1 (1877):778–800.

[18]Leonard Eugene Dickson, *History of the Theory of Numbers*, 3 vols. (New York: G. E. Steckert and Company, 1934), 1:170.

[19]H. W. Lloyd Tanner, "An Arithmetical Theorem," *Messenger of Mathematics* 7 (1878):63–64.

[20]Dickson, *History of the Theory of Numbers*, 1:340.

[21]Charles Berdellé, "La numération binaire et la numération octavale," *Association francaise pour l'avancement des sciences* (1887):206–209.

[22]Oskar Simony, "Über den Zusammenhang gewisser topologischer Thatsachen mit neuen Sätzen der höheren Arithmetik und dessen theoretische Bedeutung," *Sitzungsberichte der Mathematisch-Naturwissenschaftlichen Classe der Kaiserlichen Akademie der Wissenschaften, Wien* 96(II) (1887): 191–286.

[23]Dickson, *History of the Theory of Numbers*, 1:398.

[24]F. J. Studnicke, "Über eine neue Eigenschaft von Zahlen in 2n-zifferigen Systemen," *Sitzungsberichte der Königlichen Bömischen Gesselschaft der Wissenschaften* 7 (1896):7–10.

[25]Robert M. Pierce, *Problems of Number and Measure*, (Chicago: Robert M. Pierce, 1898).

[26]Giuseppe Peano, *Formulaire de mathématiques*, 3 vols. (Paris: Gauthier-Villars et Fils, 1901), 3:54–55.

[27]Alfred Taylor, "Octonary Numeration," *Proceedings of the American Philosophical Society* 24 (1887):296–366.

[28]William Woolsey Johnson, "Octonary Numeration," *Bulletin of the New York Mathematical Society* 1 (1891):5, 6.

[29]E. Gelin, "Du meilleur système de numération et de poids et mesures," *Mathesis, Recueil Mathématique* (2)6 (1896):161–64.

[30]T. N. Thiele, "Quel nombre serait à préfére comme base de notre système de numération?" *Danske Videnskarbenes Selskabs, Oversigt over det* (1889):25–42.

[31]Herbert Spencer, "Against the Metric System," *Appleton's Popular Science Monthly* (June 1896):199.

·VI·

THE TWENTIETH CENTURY UP TO THE COMPUTER AGE

Moritz Cantor (1901)

The publications dealing with numeration systems that have already been reviewed did not loom large in Moritz Cantor's eyes. His monumental 1901, 4-volume work on the history of mathematics found room for only minor comments, such as have already been cited in the assessments of Caramuel and Weigel. Indeed, aside from these comments, only Leibniz's dyadics seems to have found some room in this work. At the beginning of Chapter 97—which deals with series—Cantor stated:

> We will permit ourselves, at this opportunity, to mention a few investigations, for which it is doubtful that they have a right to be treated here, but which fit less well anyplace else.[1]

Cantor then proceeded to mention Leibniz's dyadics.

Thus, up to 1900, the topic of binary and other nondecimal numeration had neither blossomed into great significance nor withered away entirely, and this continued to be the story for the first half of the twentieth century, when a minor, continued interest kept the topic alive—especially in the area of recreational mathematics.

Binary Numeration Applied to the Game of Nim

Leibniz had encouraged analysis of the columns created by making vertical lists of some type of number, say squares or cubes, in binary notation—in search for columnar periods. The only advantage in knowing such periods ever pointed out by Leibniz or any of his contemporaries was the greater ease with which one could quickly write such lists on subsequent occasions. After Dangicourt's 1710 article, which had carried this fruitless analysis to boring extremes, no other writer seemed to have bothered. Charles L. Bouton (1901–1902) gave columnar analysis a new twist, however.[2] He was not concerned with columnar periods—his lists were too short for that—but he found some significance in columnar *sums*.

The significance of these columnar sums is connected with a game which Bouton called NIM. The game is played by two players P and Q who take turns removing matches from a table; the winner is the one who removes the last match. At the outset, the table contains an arbitrary number of piles and each pile contains an arbitrary number of matches. At each turn, a player chooses a single pile from which to remove one or more (possibly all) matches. If at the completion of P's turn, the number of matches are a, b, c ... in piles A, B, C ..., and the numbers a, b, c, ... are listed vertically in binary notation, then player P has a *winning* position, according to one of Bouton's theorems, if, and only if, each columnar sum of that list is an even number.

Among the lists shown below, II, IV, VI, VIII, and X indicate winning positions:

	I	II	III	IV	V	VI	VII	VIII	IX	X
a	0010	0010	0010	0010	0010	0010	0001	0001	0001	0000
b	0011	0011	0011	0011	0011	0010	0010	0001	0000	0000
c	0110	0110	0110	0001	0000	0000	0000	0000	0000	0000
d	1111	0111	0000	0000	0000	0000	0000	0000	0000	0000
	1242	0242	0131	0022	0021	0020	0011	0002	0001	0000

In fact, these ten lists represent a sample game of NIM, won by P. List I is the position at the outset, there being four piles A, B, C, and D having respectively 2, 3, 6, and 15 matches. Player P started by removing 8 matches from D, thus creating the winning position shown in II, where every columnar sum is even. None of the choices open to player Q could now have created a winning position. Q chose to remove the remaining

7 matches from D. Now P removed 5 matches from C, leaving the winning position IV and so on.

Other writers dealt with this topic subsequently. W. A. Wythoff (1907) proposed a modification of NIM limited to two piles.[3] The accompanying mathematical theory, surprisingly, involved the greatest integer function and the expression $\frac{1}{2}(\sqrt{5} + 1)$.

E. H. Moore (1910) proposed a modification, which permitted each player to remove matches from k piles.[4] He showed that this modification also had a complete mathematical theory.

R. Sprague (1935–36) saw Bouton's NIM as a special case of a more generalized NIM whose theory he presented.[5]

E. U. Condon (1942) described a machine, the NIMATRON, "which is very skillful at playing the game of NIM."[6]

D. P. McIntyre (1942) presented a new mathematical theory of NIM based on the base 4 representation of numbers.[7]

H. D. Grossman (1945) showed that a mathematical theory is also possible when NIM is modified so that the object is to make one's opponent finish with an odd or even total.[8]

L. S. Recht (1943) generalized McIntyre's treatment by showing a proof based on number representations to the base 2^q.[9]

Raymond Redheffer (1948) described a new machine for playing NIM that is an improvement over Condon's.[10] He noted that the theory of the game did not really require knowledge of the sums of columns, but only whether such sums were odd or even. The circuits for his machine were simplified accordingly.

Cedric Smith (1968) treated NIM as a special case of a more generalized game.[11]

Three books (as distinct from the journal articles reviewed up to now), each in English and widely available, can be especially recommended for their accounts of the game of NIM and the relevant mathematical theory—at least up to Bouton's work. These books are by G. H. Hardy, J. V. Uspensky, and W. W. R. Ball.[12]

The Role of Numeration Systems in Books on Recreational Mathematics and Books on the Theory of Numbers

The first edition of Wilhelm Ahrens' book on recreational mathematics appeared in 1901 and was followed by a second edition in 1910. Each contained a long chapter on numeration systems. The earlier one mistakenly referred to Simon Stevin and Charles XII of Sweden as duodecimal advocates, but the later edition retracted these references. The chapter in the 1910 edition was entitled "Numerationsysteme," and covered pages 24–104. It easily represented the most comprehensive treatment of numeration systems and their applications up to that time.

Ahrens gave concise historical sketches, provided flawless bibliographic references, and gave careful proofs of theorems. He gave credit to Lucas for applying numeration systems to the games of The Tower of Hanoi and the Chinese Rings.

Paul Bachman's book (1902) on the theory of numbers discussed briefly the representation of numbers in various "numeral systems" (Ziffernsystemen).[13] It mentioned that

$$(1012032)_7 = (120759)_{10}$$

and that a base β representation is possible for every number and every base β. He pointed out that the powers of 3 are the only ones that can be used in conjunction with the coefficients -1, 0, $+1$ to represent every number uniquely. Bachman did not include any applications, such as the Guess the Number Cards or Weighing Problems.

The Hardy book, already mentioned as a good reference on the game of NIM, followed the Barlow tradition more clearly. Chapter IX, "The Representation of Numbers by Decimals," includes all of Barlow's 1811 treatment in concise form, plus a thorough treatment of fractional numbers and their β-adic representations as well as the game of NIM. Hardy's use of the word "decimal" requires some comment. Hardy referred to $(0.001)_2$ as a "'binary' decimal" and to $(0.44\bar{4}...)_7$ as a "decimal in the scale of 7." In a footnote, however, he commented: "We ignore the verbal contradiction involved in the use of 'decimal'; there is no other convenient word." Hardy's Theorem 136, quoted below, further illustrates his use of 'decimal.'

> Suppose that r is a prime or a product of different primes. Then any positive number ζ may be represented uniquely as a decimal in the scale of r. An infinity of the digits of the decimal are less than $r - 1$; with this reservation, the correspondence between the numbers and the decimals is (1,1).[14]

Few other writers seem to have decided on this usage of 'decimal.' One of them was Don Pedoe, who referred to 101 (the binary equivalent of 5) as a "binary decimal."[15]

Edmund Landau's book (1927) on the theory of numbers omitted the topic of numeration systems altogether.[16]

Uspensky's book has already been cited as a good reference on NIM. Aside from this, numeration systems are covered much less thoroughly than in Hardy.

The 1938 Ball & Coxeter edition of *Mathematical Recreations and Essays* does not treat numeration systems as comprehensively as does Ahrens, nor is everything on this topic collected into a single section or chapter. However, its treatment of several subtopics, such as NIM and the weighing problems, is more extensive than Hardy's. This makes Ball & Coxeter one of the best precomputer age references on numeration systems in the English language.

Miscellaneous Publications on Numeration Systems

John Tennant (1901) described how he could facilitate the search for factors in high numbers N by representing them in the form $N = a(900^2) + b(900) + c$, that is, in base 900.[17]

Adam Wizel (1904) reported the case of an imbecile with unusual computational talent that involved use of base 16.[18] When asked how many buttons were in her collection of 104 buttons, the imbecile would answer "6 times 16 and 8 more." Wizel found other evidence that the imbecile's computational work was oriented around grouping in 16s.

Allan Cunningham (1908) investigated "binal fractions," the name he gave to reciprocals of (1/n) of any number n, when "expressed in the binary scale."[19] He noted that

$$1/2 = .1$$
$$1/(2^2) = .01$$
$$1/(2^3) = .001$$

and in general that

$$1/(2^a) = .000\ldots01$$

with (a − 1) zeros before the terminal 1. He noted further that if n is odd, then the 'binal' of 1/n would consist solely of a repeating cycle, as for example,

$$1/3 = .0101\overline{01}\ldots\ .$$

If, on the other hand, n were even (other than a power of 2) then the repeating cycle in the binal of 1/n would be preceeded by a group of zeros, as in

$$1/6 = .0010101\overline{01}\ldots$$

or

$$1/12 = .00010101\overline{01}\ldots\ .$$

One might expect this from the fact that 1/6 = (1/2)(1/3) or, in binals, $1/6 = (.1)(.0101\overline{01}\ldots)$ and similarly, 1/12 = (1/4)(1/3). Hence Cunningham decided to restrict his tables of binals of 1/n to those n that are odd.

Joseph Bowden (1912) proved the validity of the ancient Russian peasant method of multiplication, which he showed to be an application of the binary system.[20]

G. H. Cooper (1920) urged that "the government should appoint a commission to test the octonary, or any other system, and give the world that one which proves itself to be universally and conveniently adapted to the needs of mechanical science and trade."[21]

Alfred Watkins (1920) urged adoption of the octal system at least for fractional numbers, and designed an "octaval" ruler, simple calipers, simple vernier calipers, and double vernier calipers that show octal divisions of the inch.[22]

Louis-Gustave Dupasquier (1921) thoroughly discussed "which base is best," and concluded that from the point of view of divisibility properties some multiple of 6 would make the best base.[23] From various points of view, however, base 4 had the most advantages.

E. M. Tingley (1934) expressed strong feelings against the decimal system and against more complete adoption of the metric system. He chided psychologists for not joining his crusade. He stated: "Psychologists, that vast influential and international group, students of the mind, should assert in their authority that eight is the best base for arithmetic."[24] On this, E. William Phillips (1936) wrote:

> 1. The development of this paper involves the consideration of certain arithmetical, mechanical, and photo-electrical technicalities, but the ultimate aim is a very simple one and it may therefore be advisable to state it at once. The ultimate aim is to persuade the whole civilized world to abandon decimal numeration and to use octonal numeration in its place; to discontinue counting in tens and to count in eights instead.
>
> 2. However, it seems unlikely that the whole civilized world will be persuaded to complete this change during the next twelve months, having previously declined similar invitations. Therefore the more immediate aim is the adoption of octonal numeration for scientific and business purposes, for the great mass of figures recorded and manipulated for the benefit only of the scientific and business man, the few final results required for presentation to the layman being transformed into the denary scale of notation from the octonary by means of conversion tables, or otherwise.[25]

Joseph Bowden published his *Special Topics in Theoretical Arithmetic* in 1936. (A journal article of Bowden's has already been mentioned for 1912). In his new book he devoted pages 17–81 to "Scales of Notation." He wished to abolish the "tyranny of ten" and have, among other things, a 16-hour day.

A. J. Kempner (1936) investigated certain nonstandard numeration systems, because a student of his had asked whether the irrational number e could be used as the base of a numeration system. He answered in the affirmative, by proving

Theorem 6.0: *Every real number can be represented canonically to any base β, where β is any real number greater than 1, and this representation is unique.*

For $\beta = 3/2$, for example,

$$2 = (10.01000001\ldots)_{3/2} \text{ (can)}$$
$$2 = (\ 0.111\ldots\ )_{3/2}$$

Only the former is called *canonical*, since it satisfies the following definition.

The right side of

$$N = (D_n D_{n-1} \ldots D_0 D_{-1} D_{-2} \ldots)_\beta \tag{6.1}$$

is called the *canonical* base β representation of N if

(i) each digit D_i is a nonnegative integer less than β
(ii) $\beta^n \leqq N < \beta^{n+1}$
(iii) the digits D_i are chosen in the order $i = n$, $n - 1, \ldots$ and each digit is the highest choice possible up to that point.

If β lies in the interval (1,2), then only the two digits 0 and 1 are required, as was shown for $\beta = 3/2$. By Kempner's definition, if (in 6.1)) $D_{-1} = D_{-2} = \ldots = 0$, then N is represented as a *whole number* to the base β. The base 3/2 representations (canonical and otherwise) already shown, reveal that the integer 2 does not have a 'whole number' representation in that base and the fact that

$$5/2 = (11)_{3/2}$$

brings forth that a 'whole number' need not represent an integer. In Kempner's words:

> On account of these properties, the notion of a 'whole number' cannot be fundamental in such number systems, since it possesses no closure properties. However, it was obvious from the start that only the representation of the individual number in the system, and the abstract structure of the system, can interest us, not its use for arithmetic.[26]

Among the many other results proved or indicated by Kempner are the following:

Theorem 6.2: *If* $\beta = 3/2$, *the canonical representations exclude the sequence* 11, *the sequence* 101 *is followed by at least* 5 *zeros, and every infinite representation has a preponderance of* 0's—*the ratio of* 1's *to* 0's *being at most* 1/2.

Theorem 6.3: *If* β *is not integral, an unending repetition of the greatest digit,* $[\beta]$, *cannot occur in the base* β *canonical representation.*

Theorem 6.4: *No irrational number can be periodic to a rational base* β, *canonical or otherwise.*

Theorem 6.5: *If* $\beta = 5^{\frac{1}{2}}$ *the canonical representation cannot have the sequences* 22 *or* 21; *the following sequences of three digits are also excluded*: 222, 221, 220, 212, 211, 210, 202, 122.

As an example of a canonical representation to an irrational base, Kempner gave the example

$$1/2 = (0.100101\ldots)_\beta,$$

where $\beta = 5^{\frac{1}{2}}$. He arrived at this through the following steps:

$$\begin{aligned}
1/2 &= 1\cdot 5^{-1/2} + (1/2 - 5^{-1/2}) \\
&= 1\cdot 5^{-1/2} + 0\cdot 5^{-1} + (1/2 - 5^{-1/2}) \\
&= 1\cdot 5^{-1/2} + 0\cdot 5^{-1} + 0\cdot 5^{-3/2} + (1/2 - 5^{-1/2}) \\
&= 1\cdot 5^{-1/2} + 0\cdot 5^{-1} + 0\cdot 5^{-3/2} + 1\cdot 5^{-2} + (23/50 - 5^{-1/2}) \\
&= 1\cdot 5^{-1/2} + 0\cdot 5^{-1} + 0\cdot 5^{-3/2} + 1\cdot 5^{-2} + 0\cdot 5^{-5/2} + 1\cdot 5^{-3} + \ldots \\
&= (100101\ldots)_\beta, \text{ where } \beta = \sqrt{5}\ .
\end{aligned}$$

In a footnote, Kempner indicated that 0 and 1 were clearly excluded as possible bases, but positive numbers less than 1 and negative numbers could be used with slight modifications of the processes outlined and with suitable restrictions on the set of digits employed.

G. W. Wishard (1937) chided Taylor for having overlooked

the beneficial ease with which base 8 may be converted to base 2, thus making available the additional advantages of base 2.[27] Wishard gave an algebraic proof of the fact that a binary-coded octal string (created by substituting 000, 001, ... , 111 for 0, 1, ... , 7 respectively) is at once a pure binary string, which, of course, amounts only to the old Legendre insight.

Abraham Fraenkel (1939) preferred bases 6 or 12 to 10 for the usual reasons (richness in divisors). He added that this same principle of preference has resulted in 24 hours in a day, 60 minutes in an hour, 360 degrees in a circle, and 1080 "parts" to an hour in the Jewish calendar.

> From a purely scientific point of view, preference must be given to that number, among the infinite possibilities of choice for a base, which is absolutely distinguished from the rest as the smallest among them, namely, the number 2. As a matter of fact, in so far as positional notation is employed in purely mathematical investigations, the binary scale . . . is sometimes chosen.[28]

J. T. Johnson (1940), of the Chicago Teachers College, reminded his readers of the difficulties still being experienced in trying to get the metric system fully adopted throughout the world, and thought it futile for anyone to suggest the more far-reaching reform of replacing 10 as the base of our arithmetic. He hoped for *full* adoption of the metric system in the United States, a reform "that no one can deny is in the interest in simplicity for education, computation, business and commerce." Johnson analyzed Tingley's intemperate advocacy of base 8 and concluded:

> All the discussions on number systems besides our own have their value, the chief one of these is a development of a mathematical appreciation which is so much underrated and neglected at the present time. Another value in these discussions lies in the material they furnish for the imagination. When we realize that more than one half of the English books that are written and read are fiction rather than fact, there ought to be no objection to more books like *New Numbers* by

Mr. Andrews. They belong to mathematical fiction and should have a place in mathematical literature.[29]

Richard Courant (1941) included an exercise on page 9 of his *What is Mathematics*. The exercise, in effect, asked his readers to work out a table like Hankel's (see chapter 5), showing base 4 to be 'best', because it requires the least number of concepts or names.

Duncan Harkin (1941) gave a justification for the 'repeated division by 2' method of finding binary equivalents of N that seems unusually well suited to his intended readership.[30] He also gave most of the 'recreational' applications of nondecimal numeration (NIM, weighing problems, peasant multiplication), and may be forgiven a hopelessly distorted reference to Lagny's work.

Harold Larsen (1944) reported that if the base be taken as $1 + i$, where i is a positive number less than unity, and where 0 and 1 are the only permissible digits, then one could see some interesting relationships to the mathematics of finance.[31] He indicated that circulating 'decimals' in the scale of $1 + i$ are connected with the theory of perpetuities and that certain logarithms could then be identified with the sinking fund and amortization schedules.

J. Ser (1944) proposed a nonstandard numeration system whose digit strings would readily expose the divisibility properties of N with respect to certain prime moduli.[32] Thus, for example,

$$24 = 0043 \quad \text{(ser-base 210)}$$

and the 4-digit string 0043 would mean that 24 is a multiple of 2 and 3, and exceeds a multiple of 5 by 4, and a multiple of 7 by 3. The prime factorization of 210 being $2 \cdot 3 \cdot 5 \cdot 7$ would indicate that the primes 2, 3, 5, and 7 are being considered in turn. Similarly,

$$174 = 0046 \quad \text{(ser-base 210)}$$

because 174 exceeds multiples of 2, 3, 5, and 7 by 0, 0, 4, and 6 respectively. Since 30 has only the prime factors 2, 3, and 5, it follows that

$$24 = 004 \quad \text{(ser-base 30)}$$

L. R. Posey (1946) reminded his readers that the formula

$$\mathrm{Log}_a N = \frac{\mathrm{Log}_b N}{\mathrm{Log}_b a}$$

intended for a change of base in logarithms, is intimately connected with changes in base in our numeration system.[33]

Duodecimal Advocates (1900–1946)

Scientific American reported in 1902:

> The American Society of Mechanical Engineers has apparently not yet given up the idea of combating the introduction of the metric system into the United States. As a kind of compromise between the existing system and the metric system, Prof. S. S. Reeve recently proposed before the Society a duodecimal system, which takes as its standard the English yard. Upon the yard a system is to be reared, exactly as a system has been built up upon the meter. The divisions, however, are duodecimal to suit duodecimal numbers.

L. H. Vincent (1909) advocated that his readers take a good look at base 12, but admitted the hopelessness of ever replacing base 10.[34]

R. P. William (1909) came to the same conclusion as Vincent after giving an extensive historical overview for base 12.[35]

H. C. Christofferson (1924) also agreed, after his review of some of the advantages of base 12.[36]

F. Emerson Andrews (1934, 39, 44) wrote extensively on the advantages of base 12. In one of his works he stated:

> The lowest number that has four factors is 12; the lowest that has six factors, two times 12; the lowest that has seven factors, three times 12; the lowest that has eight factors, four times 12; the lowest that has ten factors, five times 12; all the others that have the maximum of ten factors, excepting only 90 are six, seven, and eight times 12.[37]

In the end, his three publications offer only 'richness in

divisors' and 'resulting simplicity of fractions' as the advantages of base 12, although in considerable detail.

Wilimina Pitcher (1934) published a play intended for pupil use concerning a dozenland in which people had 12 fingers and used base 12.[38]

Luise Lange (1936) proposed a system of names to go with the duodecimal system, but urged her readers to be realistic and forget about ever replacing base 10 for common use.[39]

James Johnston (1938) applied the Cauchy modification (see chapter 5) to the duodecimal system.[40]

George Terry (1938) pointed out that in the duodecimal system, all primes end in 1, 5, 7, or B and all squares in 0, 1, 4, or 9.[41]

L. C. Janes (1944) echoed some of the advantages of base 12 that had previously been pointed out and concluded:

> In closing this discussion, we venture to suggest that if a committee were ever appointed to study the possibility of a uniform system of weights and measures to be used throughout the entire world with a view to the adoption of a system that would be nicely adaptable to decimal computation it might be well to investigate the possibilities of the duodecimal system[42]

In 1944 the Duodecimal Society of America announced its formation. According to its constitution the Society existed "to conduct research and education of the public in mathematical science, with particular relation to the use of Base Twelve in numeration, mathematics, weights and measures, and other branches of pure and applied science."[43] During 1980 this Society was renamed DOZENAL SOCIETY OF AMERICA. Its library and permanent headquarters is now located c/o Math Department, Nassau Community College, Garden City, NY 11530.

Nondecimal Numeration in Textbooks (1900–1946)

In 1925 Gordon Mirick and Vera Sanford reported that "like other interesting by-paths in mathematics, this topic [Scales of Notation] appears only rarely in text books."[44]

Among the three books they mentioned, only one was in English, namely: Siceloff and Smith, *College Algebra* (Ginn, 1924:pp.239–246).

To this list might be added George Chrystal's *An Elementary Textbook of Algebra*. The 5th edition of this work appeared in 1904 and served to educate several generations of English mathematicians. Chapter IX of volume I is headed "Further Applications to the Theory of Numbers—On Various Ways of Representing Integral and Fractional Numbers." Here may be found a rather thorough coverage of β-adic notation, including the Cauchy modification. The more important divisibility theorems (such as by $\beta \pm 1$), are also given, but applications to NIM and other recreational mathematics are not. The 43 problems on this topic exceed in variety and difficulty those found in most theory of number books. The reader is asked to prove, for example, that "In the scale of 11 every integer which is a perfect 5th power ends in one or other of the three digits 0, 1, a," and "If in the scale of 12 a square integer (not a multiple of 12) ends with 0, the preceding digit is 3, and the cube of the square root ends with 60."

Chrystal's work (p. 215, vol. I) also shows an application of the binary system to the finding of surds of the type $p^{1/n}$ that requires $1/n$ be written in binary notation.[45] The procedure involves repeated square roots and is reminiscent of Legendre's application of repeated squaring for finding p^n. Chrystal admits that this method, "although interesting in theory, would be very troublesome in practice." A *seventh* edition of Chrystal appeared in 1964.

Published after Mirick and Sanford's lament about the scarcity of textbooks that cover this topic, Herbert Buchanan's book appeared with an appendix on "Scales of Notation."[46] Except for this heading the treatment might have been lifted from a secondary school textbook of the 1960s. It included, for example, careful use of expanded notation to show that our ordinary arithmetic deals with the "detached coefficients" of polynomials in 10.

Franklin Kokomoor's book, intended for college freshmen,

did not tuck this topic into an appendix, but smoothly integrated it in a discussion of number representations that included ancient numeration.[47] Neither of these last two books went as deeply into the topic as Chrystal. Kokomoor's and Chrystal's books devoted respectively 1% and 1.5% of their pages to nondecimal numeration.

NOTES TO CHAPTER VI

[1]Moritz Cantor, *Vorlesungen über Geschichte der Mathematik*, 4 vols. (Leipzig: B. G. Teubner, 1922), 3:360–61.

[2]C. L. Bouton, "Nim: A Game with a Complete Mathematical Theory," *Annals of Mathematics*, series 2, 3 (1901–1902):35–39.

[3]W. A. Wythoff, "A Modification of the Game of Nim," *Nieuw Archief voor Wiskunde* 7 (1907):199–202.

[4]E. H. Moore, "Nim," *Annals of Mathematics*, series 2, 11 (1910):93–94.

[5]R. Sprague, "Über mathematische Kampfspiele," *Tohoku Mathematical Journal* 41 (1935–1936):438–444.

[6]E. U. Condon, "The Nimatron," *American Mathematical Monthly* 49 (May 1942):330–32.

[7]D. P. McIntire, "A New System for Playing the Game of Nim," *American Mathematical Monthly* 49 (January 1942):44–46.

[8]H. D. Grossman and D. Kramer, "A New Match Game," *American Mathematical Monthly* 52 (October 1945):441–43.

[9]L. S. Recht, "The Game of Nim," *American Mathematical Monthly* 50 (August-September 1943):435.

[10]Raymond Redheffer, "A Machine for Playing the Game of Nim," *American Mathematical Monthly* 55 (June-July 1948):343–49.

[11]Cedric A. B. Smith, "Compound Games with Counters," *Journal of Recreational Mathematics* 1 (April 1968):67–77.

[12]G. H. Hardy and E. M. Wright, *An Introduction to the Theory of Numbers*, 3rd ed., (Oxford: Clarendon Press, 1954), pp. 117–120; J. V. Uspensky and M. A. Heaslet, *Elementary Number Theory*, (New York: McGraw-Hill, 1939), pp. 16–19; W. W. Rouse Ball and H. S. M. Coxeter, *Mathematical Recreations and Essays*, (New York: Macmillan, 1960), pp. 36–38.

[13]Paul Bachman, *Niedere Zahlentheorie*, (Leipzig: B. G. Teubner, 1902).

[14]Hardy and Wright, *An Introduction to the Theory of Numbers*, p. 112.

[15]Dan Pedoe, *The Gentle Art of Mathematics*, (New York: Macmillan, 1958), p. 17.

[16]Edmund Landau, *Vorlesungen über Zahlentheorie—Aus der elemtaren Zahlentheorie*, (New York: Chelsea Publishing Company, 1950).

[17]John Tennant, "On the Factorisation of High Numbers," *Quarterly Journal of Pure and Applied Mathematics* 32 (1901):322–42.

[18]Adam Wizel, "Ein Fall von phänomenalem Rechentalent bei einem Imbecillen," *Archiv für Psychiatrie und Nervenkrankheiten* 38 (1904):122–55.

[19]Allan Cunningham, "On Binal Fractions," *The Mathematical Gazette* 4 (1908):259–67.

[20]Joseph Bowden, "The Russian Peasant Method of Multiplication," *The Mathematics Teacher* 5 (1912):4–8.

[21]G. H. Cooper, "Eight-digit System," *Scientific American* 122 (January 1920:111.

[22]Alfred Watkins, "Octaval Notation and the Measurement of Binary Inch Fractions," *American Machinist* 52 (March 1920):685–88.

[23]Louis-Gustave DuPasquier, *Le développement de la notion de nombre, (Paris: Attinger Frères, 1921).*

[24]E. M. Tingley, "Calculate by Eights, Not by Tens," *School Science and Mathematics* 34 (April 1934):395–99.

[25]E. William Phillips, "Binary Calculation," *Journal of the Institute of Actuaries* 67 (1936):187–221.

[26]A. J. Kempner, "Abnormal Systems of Numeration," *American Mathematical Monthly* 43 (December 1936):610–17.

[27]G. W. Wishard, "The Octo-binary System," *National Mathematics Magazine* 11 (March 1937):253–54.

[28]Abraham Adolf Fraenkel, "Natural Numbers as Cardinals," *Scripta Mathematica* 6 (June 1939):69–79.

[29]J. T. Johnson, "New Number Systems vs the Decimal System," *School Science and Mathematics* 40 (December 1940):828–34.

[30]Duncan Harkin, *Fundamental Mathematics*, (New York: Prentice-Hall, 1941).

[31]Harold Larsen, "A Note on Scales of Notation," *American Mathematical Monthly* 51 (May 1944):274–75.

[32]J. Ser, *La Numération et le Calcul des Nombres*, (Paris: Gauthier-Villars et Fils, 1944).

[33]L. R, Posey, "Change of Base," *School Science and Mathematics* 46 (December 1946):871–78.

[34]L. H. Vincent, "Duodecimal System of Notation," *School Science and Mathematics* 9 (June 1909):555–62.

[35]R. P. William, "Ancient Duodecimal System," *School Science and Mathematics* 9 (June 1909):516–21.

[36]H. C. Christofferson, "A New Number System," *School Science and Mathematics* 24 (December 1924):913–16.

[37]F. Emerson Andrews, *New Numbers*, (New York: Essential Books, 1944), p. 46.

[38]Wilimina E. Pitcher, "Alice in Dozenland," *The Mathematics Teacher* 27 (December 1934):390–96.

[39]Luise Lange, "On Fingerprints in Number Words," *School Science and Mathematics* 36 (January 1936):13–19.

[40]James Halcro Johnston, *The Reverse Notation—Introducing Negative Digits with Twelve as Base*, (London: Blackie and Son, 1938).

[41]George S. Terry, *Duodecimal Arithmetic*, (London: Longmans, Green and Company, 1938).

[42]W. C. Janes, "The Duodecimal System," *The Mathematics Teacher* 37 (December 1944):365–67.

[43]"The Duodecimal Society of America," *School Science and Mathematics* 44 (November 1944):694. During 1980 this Society was renamed Dozenal Society of America. Its library and permanent headquarters is now located c/o Math Department, Nassau Community College, Garden City, NY 11530.

[44]Gordon R. Mirick and Vera Sanford, "Scales of Notation," *The Mathematics Teacher* 18 (1925):465–71.

[45]George Chrystal, *An Elementary Textbook of Algebra*, 2 vols. (Edinburgh: A. and C. Black, 1919–20), 1:215.

[46]Herbert E. Buchanan et al., *A Brief Course in Advanced Algebra*, (Boston: Houghton Mifflin Company, 1937).

[47]Franklin Wesley Kokomoor, *Mathematics in Human Affairs*, (New York: Prentice-Hall, 1942).

·VII·

APPLICATIONS TO COMPUTERS

UP to 1946 only mathematicians and an occasional king, science adviser, engineer, teacher, or other such personage seemed to be aware of the existence of nondecimal numeration. This situation changed as the progressing age of electronic computers made even the man in the street aware of the binary system. Ironically, this age began with the appearance of the ENIAC (Electronic Numerical Integrator and Computer), the most decimal of 'decimal' computers.

NUMBER REPRESENTATION IN THE ENIAC

John W. Mauchly had conceived of this computer in 1942, though the project of building it involved at least 30 other engineers and mathematicians.[1] Public knowledge of ENIAC was delayed to February 14, 1946 because of wartime secrecy. Internally, ENIAC reserved 10 flip-flop switches for each decimal digit, and only one of these switches could be in the ON position.[2] Corresponding external lights reflected the state of those switches. Since there were 10 banks of 10 lights each, integers as high as 9 999 999 999 could be represented. The number 1984, for example, would in effect appear as follows:

1984 = (00000 00010, 10000 00000, 01000 00000, 00000 10000).

In effect, each decimal digit was assigned a 'one out of ten' binary string as indicated below:

0	00000 00001
1	00000 00010
2	00000 00100
3	00000 01000

4	00000 10000
5	00001 00000
6	00010 00000
7	00100 00000
8	01000 00000
9	10000 00000

To contrast with this ENIAC number representation, 1984 is given below in BCD (binary coded decimal) and pure binary notation:

(7.1) 1984 = (0001 1001 1000 0100) (BCD)
(7.2) 1984 = (11111000000) (pure binary)

It might be argued that a computer that uses 7.1 is decimal, but it also uses the binary system in a minor way to represent each decimal digit. If it is argued that this also applied to the ENIAC number representation, then even an automobile's odometer could be said to be 'binary,' since each of the ten positions on one of its wheels is in one of two 'binary' states, namely 'hidden' or 'visible.'

The Burks-Goldstine-Von Neumann Report

The special suitability of true binary numeration for computers was brought out strongly in this 1947 report.

> In a discussion of the arithmetical organs of a computing machine one is naturally led to a consideration of the number system to be adopted. In spite of the longstanding tradition of building digital machines in the decimal system, we feel strongly in favor of the binary system for our device. Our fundamental unit of memory is naturally adapted to the binary system since we do not attempt to measure gradations of charge at a particular point in the Selectron, but are content to distinguish two states. The flip-flop again is truly a binary device. On magnetic wires or tapes and in acoustic delay line memories one is also content to recognize the presence or absence of a pulse or (if a carrier frequency is used) of a pulse train, or of the sign of a pulse. Hence, if one contemplates using a decimal system with

either the Iconoscope or delay line memory one is forced into a binary coding of the decimal system—each decimal digit being represented by at least a tetrad of binary digits. Thus an accuracy of ten decimal digits requires at least 40 binary digits. In a true binary representation of numbers, however, about 33 digits suffice to achieve a precision of 10^{10}. The use of the binary system is therefore somewhat more economical of equipment than is the decimal.

The main virtue of the binary system as against the decimal is, however, the greater simplicity and speed with which the elementary operations can be performed. To illustrate, consider multiplication by repeated addition. In binary multiplication the product of a particular digit of the multiplier by the multiplicand is either the multiplicand or null according as the multiplier digit is 1 or 0. In the decimal system, however, this product has ten possible values between null and nine times the multiplicand, inclusive. Of course, a decimal number has only $log_{10}2 \simeq 0.3$ times as many digits as a binary number of the same accuracy, but longer than in the binary system. One can accelerate decimal multiplication by complicating the circuits, but this fact is irrelevant to the point made since binary multiplication can likewise be accelerated by adding to the equipment. Similar remarks may be made about other operations.

An additional point that deserves emphasis is this: An important part of the machine is not arithmetical, but logical in nature. Now logics, being a yes-no system, is fundamentally binary. Therefore, a binary arrangement of the arithmetical organs contributes very significantly towards a more homogenous machine, which can be better integrated and is more efficient.

The only disadvantage of the binary system from the human point of view is the conversion problem. Since, however, it is completely known how to convert numbers from one base to another and since this conversion can be effected solely by the use of the usual arithmetic processes there is no reason why the computer itself cannot carry out this conversion. It might be argued that this is a time consuming operation. This, however, is not the case . . . . Indeed a general purpose computer, used as a scientific research tool, is called upon to do a very great number of multiplications upon a relatively small amount of input data, and hence the time consumed in the decimal to binary conversion is only a trivial percent of the total computing time. A similar remark is applicable to the output data.

In the preceding discussion we have tacitly assumed the desirability of introducing and withdrawing data in the decimal system. We feel, however, that the base 10 may not even be a permanent feature in a scientific instrument and consequently will probably attempt to train ourselves to use numbers base 2 or 8 or 16. The reason for base 8 or 16 is this: Since 8 or 16 are powers of 2 the conversion to binary is trivial. Since both are about the size of 10, they violate many of our habits less badly than base 2.[3]

This report was made specifically to guide the design of a 'scientific' computer contemplated for the Institute for Advanced Study, Princeton, N.J. The special suitability of true binary numeration pointed out in the report rested in part on the assumption that the computer would be *scientific*, that is, would do a relatively large amount of computing on a small amount of data.

Having a computer act on payroll information (hours worked, pay rate, tax rate, etc.) for 5000 employees and prepare 5000 paychecks would be an example of a nonscientific use of computers, that is, data processing or business use. For such business computers, the justification for true binary numeration seemed less compelling, and the next two decades continued to see both decimal and binary computers. Consistent with the use in computer literature, a computer shall be called *binary* (even though the input and output may be decimal) if number representation in its arithmetic unit is in true binary form, i.e., like 7.2. If the computer uses decimal arithmetic in its arithmetic unit, then it will be called a *decimal* computer. As the report by Burks et al. pointed out, even a decimal computer has to code each decimal digit into at least a tetrad of binary digits, i.e., a 4-bit string, as shown, for example, in 7.1. The ENIAC had used a 10-bit string.

According to Richards, the ideas of the Burks report had been generated during the development of the ENIAC and were widely disseminated during a 1946 summer session at the

University of Pennsylvania. "Very shortly thereafter, electronic computer projects started at a number of institutions."[4]

Some Pre-Eniac Milestones Toward the Age of the Electronic Computer

That important, essentially binary computer component, the Eccles-Jordan flip-flop trigger circuit became known in 1919.[5]

In 1923, Craft et al. reported on automatic telephone switching systems as follows:

> It will be noted that the method of selection is not on a decimal basis. The first selection is to choose one of five brushes on the incoming selector . . . . The next selection is by groups of 500, which is again nondecimal. This "translation," as it is called, of the number from the decimal notation, as dialed by the subscriber, into the notation as needed by the selectors, is taken care of very simply in the senders.[6]

The year 1932 marked the first use of binary numeration in fast electronic counting circuits, or what C. E. Wynn-Williams called a "Scale of Two" counter.[7]

The four years beginning with 1938 brought the development of the Atanasoff-Berry Computer at Iowa State University.

> By today's terminology, the Atanasoff-Berry computer would be known as a special-purpose computer designed for the solution of up to 30 simultaneous algebraic equations with a corresponding number of unknowns. The mathematical scheme for solving the equations was the systematic elimination of coefficients by combining pairs of equations linearly. Internally, the binary system of number representation was used although the number input, which was by means of IBM cards, was in decimal form. The machine itself performed the radix conversion with 50 binary digits being used for each number.[8]

Even though this computer was never used (some portions were incomplete when World War II interrupted the work in 1942), Richards considered it the ancestor of all electronic digital systems and reported the following interesting link with later work:

> One of the few people to study the machine in detail was Dr. John Mauchly, who at the time was on the faculty of Ursinus College in Pennsylvania. According to oral reports from Dr. Atanasoff and Dr. Mauchly, the two met at an American Association for the Advancement of Science Meeting. As a result of conversation at this meeting, Dr. Mauchly made a visit to ISU in 1941 for the specific purpose of studying the computer . . . . Dr. Mauchly is given credit for subsequently initiating the ENIAC project.[9]

According to Edmund Berkeley, Dr. George R. Stibitz introduced the "excess three" code in 1939, using it in his Complex Computer, a special purpose machine for manipulating complex numbers.[10] This code is shown below together with two others, the 8-4-2-1 and the biquinary codes.

decimal digit	*8421*	*excess three*	*biquinary 50 43210*
0	0000	0011	01 00001
1	0001	0100	01 00010
2	0010	0101	01 00100
3	0011	0110	01 01000
4	0100	0111	01 10000
5	0101	1000	10 00001
6	0110	1001	10 00010
7	0111	1010	10 00100
8	1000	1011	10 01000
9	1001	1100	10 10000

The following examples illustrate their use:

(7.3) 1984 = 0001 1001 1000 0100 (8421)
(7.4) 1984 = 0100 1100 1011 0111 (excess three)
(7.5) 1984 = 0100010 1010000 1001000 0110000 (biquinary)

7.3 is the same as 7.1, that is, the 8421 code is the same as binary coded decimal. The excess three code uses the binary equivalent of (d + 3) for each decimal digit d (hence its name). The biquinary is also known as the (5043210)-code because of the place values, or "weights" associated with these 7-bit strings. All three belong to a class of codes known as *decimal codes*, and will be discussed as a class in a separate section. Stibitz devised the biquinary code for use in his Biquinary Calculator. In today's language, this was a decimal calculator that used a 7-bit code for the decimal digits. Stibitz explained his choice of the name "biquinary" as follows:

> The present system is known as the biquinary system, the appellation being derived from the fact that each digit is expressed through the selection of one of two alternatives out of a first group of two and one of five alternatives out of a second group of five. One of the groups may thus be said to be on a binary basis whereas the other may be said to be on a quinary basis and the combination of the two groups is, therefore, said to be on a biquinary basis.[11]

It should be noted that this biquinary code involves neither radix 2 nor 5. The number 5 does appear as one of the 7 place values—put paradoxically in the portion Stibitz insisted on calling binary. Similarly, the number 2 appears in the quinary portion.[12]

Four-bit Decimal Codes

For the post-ENIAC age, it seems more appropriate to survey the variety of number representations in computers and indicate some of the reasons for this variety than to trace their appearance chronologically.

Consistent with usage in computer literature, a particular way of assigning ten n-bit strings to the ten decimal digits shall be called a *decimal code*. The ENIAC, (8421), excess-three, and biquinary codes belong to this class of codes and are 10, 4, 4, and 7 bits long respectively. A 3-bit code is not possible, there being only eight 3-bit strings from 000 to 111, but more than 29 billion (16!/6!) 4-bit codes exist (permutations of 16 things 10 at a time). The small fraction of these that have been used in computers or that are of special theoretical interest shall be surveyed in this section.

A 4-bit decimal code shall be called *weighted* if there exists a set of 4 integers (abcd) that may be interpreted as the four place values (or weights) of the ten 4-bit strings. (A similar definition holds for n-bit decimal codes where n is greater than 4.) The set of integers (abcd) shall also be known as a *weighting scheme*.

Thus, the (8421)-code is a weighted code and is in fact known by its set of weights. Can a set of weights be found for the excess-three code? Suppose (abcd) were such a set. Since in this code 0011 = 0, it would follow that $c + d = 0$. Similarly, $a = 5$, since 1000 = 5. But from this it would follow that 1011 should be assigned to 5, that is, contrary to the actual assignment. Hence, a set of weights (abcd) cannot exist for the excess-three code.

At first thought, it would appear that if true binary notation was to be avoided in favor of a decimal code, then the (8421) should do nicely. In fact, depending upon the particular application involved, designers saw various alternative codes as being better suited to their purposes. Even within the same computer, differing requirements might make more than one decimal code desirable. For example, the decimal code in use in the arithmetic unit might differ from the one in the number storage units.

The usual way of arranging for subtraction of y from x in the arithmetic unit makes it desirable that decimal codes be *self-complementing*, a term which shall be explained shortly. In the ENIAC, $x - y$ was accomplished in the manner indicated by the right side of

$$x - y = x + c(y) - 10^{10}$$

where $c(y)$ is the 10^{10}th complement of y, i.e., where

$$c(y) = 10^{10} - y.$$

The largest number that could be represented by this computer was $10^{10} - 1$, and any additions were automatically done modulo 10^{10}, since the "carry" into the (10^{10})s place was lost. For $x = 801$ and $y = 527$, to use the example reported by H. Goldstine and Adele Goldstine, $c(y) = 9\ 999\ 999\ 473$.

Given this procedure, the question arises of how easily one might arrange to find c(y) from the known y. One notes that corresponding digits of y and c(y) must add up to 9 except in the units digits, which add up to 10. Indeed, by changing the procedure slightly to

$$x - y = x + k(y) + 1 - 10^{10},$$

where

$$k(y) = (10^{10} - 1) - y$$

the exceptional case of the units digits is eliminated and k(y) can easily be found by taking the 9's complement of *each* decimal digit of y.

If the decimal digits are represented by strings of the excess-three code, then the finding of k(y) is particularly easy. Each string representing the digit d that can be changed to the string representing (9 − d) by simply reversing the value of each binary digit, i.e., replacing each "0" by "1" and each "1" by "0". For example, 1000 = 5, but 1000 becomes 0111 upon "reversal," and 0111 = 4—which is the 9's complement of 5. Such a code is called *self-complementing*.

The (8421)-code is not self-complementing, since, for example, 1000 = 8, but 0111 = 3 and 8 + 3 ≠ 9. A machine that uses the (8421) or some other decimal code that is not self-complementing must have a procedure for finding k(y) that is more cumbersome than a simple reversal of all binary digits of y.

Designers have seen two additional advantages in excess-three over the (8421)-code. One is the avoidance in the former of the 0000 string, which in the latter was indistinguishable from the complete absence of a string. The other has to do with the ease of arranging for addition of two decimal digits and making arrangements for any possible 'carry.' For example, 4 + 6 = 10 or 4 + 6 = 0 (mod 10) would temporarily involve 0111 + 1001 = 10000. The appearance of a fifth place is a simple criterion for arranging a decimal carry that is not available for the (8421)-code. However, "10000" needs a double correction for being excess-six instead of excess-three

TABLE 2

Selected 4-bit Decimal Codes and their Characteristics

Code	C1	C2	C3	C4	C5	C6	C7	C8
Weights	8421		2421	5421			7421	7421
Decimal								
0	0000	0011	0000	0000	0000	0001	0000	0000
1	0001	0100	0001	0001	0001	0010	0001	0001
2	0010	0101	0010	0010	0011	0011	0010	0010
3	0011	0110	0011	0011	0010	0100	0011	0011
4	0100	0111	0100	0100	0110	0101	0100	0100
5	0101	1000	1011	1000	0111	0110	0101	0101
6	0110	1001	1100	1001	0101	1000	0110	0110
7	0111	1010	1101	1010	0100	1001	1000	0111
8	1000	1011	1110	1011	1100	1010	1001	1001
9	1001	1100	1111	1100	1000	1100	1010	1010
Weighted?	yes	no	yes	yes	no	no	yes	yes
Self-complementing?	no	yes	yes	no	no	no	no	no
How many "1"'s?	15	20	20	15	15	16	14	16
First five easily told from last five?	no	yes	yes	yes	no	no	no	no
Avoids 0000 string?	no	yes	no	no	no	yes	no	no
Reflected?	no	no	no	no	yes	no	no	no
Error-correcting?	no	no	no	no	no	no	no	no

(the true binary sum of two excess-three strings being excess-six), the other for the decimal carry that has been made. This calls respectively for −3 and −10 or a total correction of −13, which is accomplished simply by suppressing the "1" in "10000" (which has the effect of subtracting 16) and by adding 3. After this correction, the desired result 0111 + 1001 = 0011 (mod ten) appears.

The accounts of R. K. Richards (Chapter 6) and Willis Ware indicate that computer designers look for characteristics in a decimal code that are convenient to the user of the equipment or to the designer.[13] They seek affirmative answers to one or more of the seven questions listed below. Eight examples of decimal codes and corresponding answers to the seven questions are shown in Table 2. C1 and C2 in that table are the already discussed (8421) and excess-three codes.

(1). Is it weighted? (There is at least a mnemonic advantage in weightedness.)

(2). Is it self-complementing?

(3). Does it have a minimum number of "1"'s? (Since a "1" usually indicates that some electronic device is ON, electric power use can be minimized by minimizing the use of "1"'s. By including the 0000 string and avoiding strings having 3 or 4 "1"'s, the total can be kept down to 14).

(4). Is there a simple way to distinguish the first five from the last five strings, such as having the left-most digit 0 and 1 respectively?

(5). Does it avoid the 0000 string?

(6). Is it reflected? (A code is called *reflected*, if for each decimal digit d and its successor [d + 1], the binary strings differ by only a single one of the four bits. The excess-three and [8421]-codes are not reflected, since for each, adding unity to 0111 results in 1000, requiring a total of 4 digit reversals.)

(7). Does it have error-detecting or -correcting properties?

None of the eight 4-bit codes of Table 7.1 has a YES for question 7, since this would require the longer codes discussed in the next section.

TABLE 3

Selected Decimal Codes whose Length is Greater than Four Bits

Code	C9	C10	C11	C12	C13	C14
Name	ENIAC Code	Bi-quinary				
Weights	9876543210	5043210	08421	74210	01247	63210
Decimal						
0	0000000001	0100001	10000	11000	00011	01001
1	0000000010	0100010	00001	00011	11000	00011
2	0000000100	0100100	00010	00101	10100	00101
3	0000001000	0101000	10011	00110	01100	00110
4	0000010000	0110000	00100	01001	10010	01010
5	0000100000	1000001	10101	01010	01010	01100
6	0001000000	1000010	10110	01100	00110	10001
7	0010000000	1000100	00111	10001	10001	10010
8	0100000000	1001000	01000	10010	01001	10100
9	1000000000	1010000	11001	10100	00101	11000

Weighted 4-bit codes have been studied thoroughly. It should be noted that the codes C7 and C8 have the same set of weights, namely (7421), yet the former is known as *the* (7421)-code, because it has actually been used. Some weighting schemes, such as (7421), are associated with two or more decimal codes. Others, such as (8421), define a unique code. In 1955 Richards published a list of 70 weighting schemes (abcd), each having one or more decimal codes.[14] Richards indicated that these 70 were the only ones known, but that his list might not be exhaustive since it had been found "by a cut-and-try search process." His list included 17 schemes with all weights positive; the rest had one or two weights negative.

G. P. Weeg reported in 1960 that his lemmas and associated computer searches had established, in substance:

Theorem 7.1: *There exist only* 17 (*all of which had appeared in Richards' list*) *weighting schemes with all weights positive, and of these* 17, *only one, the* (8421)-*scheme, defines a unique code, the remaining ones having two or more codes associated with themselves.*

Theorem 7.2: *There exist only* 71 (53 *of which had already appeared in Richards' list*) *weighting schemes with one or two weights negative. Of these* 71 *schemes,* 21 *define unique decimal codes.*

Theorem 7.3: *No* 4-*bit weighting scheme can exist which has more than two negative weights.*

Theorem 7.4: *In a weighting scheme* (*abcd*) *with all four weights positive, at most one weight can exceed* 4.

Decimal Codes That Are Longer Than Four Bits

Two such longer codes have already been mentioned and are repeated as C9 and C10 in Table 3. As indicated in the last section, such longer codes may be desirable for their error-detecting or error-correcting properties, although few post-ENIAC designers considered going beyond a 7-bit length for this purpose.

Even a 4-bit code has limited error-detecting properties. In the case of the (8421)-code, for example, should the string 0111 be intended and should 1111 appear erroneously instead, this could be detected, because the latter string is unassigned in this code—it is *illegal*. On the other hand, 0111 might erroneously appear as 0110, which could *not* readily be detected as an error, because 0110 is not an illegal string in that decimal code.

For every 4-bit code there are 10 4-bit strings that are legal and 6 that are not. Since there exist 2^n n-bit strings and since only 10 of these need be assigned for a decimal code, the number of illegal strings increases rapidly with n, being 6, 22, 54, and 118 for n = 4, 5, 6, and 7 respectively.

Checking the *parity* (whether the number of "1"'s is odd or even) of a string happens to be easily arranged in a computer. Hence the preference for a decimal code that has every string of the *same* parity. Such codes shall be called *parity codes*.

Code C11 of Table 3 is such a parity code. It has odd parity, that is, every string has an odd number of "1"s. Any single error in a string (a single bit having a value opposite to that intended) would be reflected in a change from odd to even parity and hence, would show up during the parity check. A double error, on the other hand, would keep parity intact and could not be detected through a parity check.

Code C11 may be thought of as the (8421)-code with a parity bit prefixed. The value of the parity bit is chosen to give each string odd parity. Since the parity bit adds no additional information and is completely determined by the other four bits, it is also known as the *redundancy* bit.

When Code C7, also known as the (7421)-code, is improved in a certain two ways, it becomes C12. One improvement, even though that destroys the weighting scheme, is the elimination of the 0000 string in favor of 1100. The second improvement is attaching a parity or redundancy bit at the right end of each string to give even parity.

Code C13 is of special interest because of its wide use in telephone dialing equipment. The essential indistinguishability of C13 and C12 becomes clear when one imagines two telephone company employees standing on either side of a set of 5 switches representing a decimal digit. The one in front will see the digit 9 in the form of 00101 in accordance with C13, the other will see the digit as 10100 in accordance with C12. Each binary digit "1" is represented by a switch that is ON and each "0" by a switch that is OFF.

This code C13 is attractive to the telephone companies because checking (by machines or human beings) is facilitated by the fact that every 5-bit string has exactly two "1"s. Such a code is called a two-out-of-five code. Similarly, the ENIAC-code might be called a one-out-of-ten code. The biquinary may be called a two-out-of-seven code or a combination code made up of a one-out-of-two and a one-out-of-five code. Moreover, which of the five switches should be ON can readily be remembered by the (01247) weighting scheme that fits all except one of the decimal digits.

When the number 1964 is dialed within a given phone exchange, that number is likely to be represented as indicated by the right side of

1964 = (11000 00101 00110 10010) (C13)

and *not* the right side of

1964 = (11110101100) (true binary),

as the 1964 Bell Telephone Company Exhibit at the New York World's Fair seemed to suggest.

Among the 32 5-bit strings from 00000 to 11111, only 10 exist that have exactly two "1"s. Hence every two-out-of-five decimal code uses some permutation of those same strings. C12, C13, and C14, for example, are permutations of the same ten strings. In general there exist $\binom{n}{2}$ two-out-of-n strings among the 2^n n-bit strings, so that again, a greater variety of codes becomes available to the designer who is willing to pay the price of additional length.

Code C14 is another example of a code that is not truly weighted. Nevertheless, the set of weights (63210) is used as a mnemonic device because it fits all but one of the strings. The last bit is a parity bit.

The methods indicated so far may identify a string that contains a single error without identifying the particular bit in error, i.e., without error-*correcting* potential. That the latter is possible at all, will be seen by the following example. Suppose the number 1984 is to be represented in (8421)-code and the information presented in matrix form as indicated by the right side of

$$1984 \quad = \quad \begin{array}{l} \mathit{1}0001 \\ \mathit{0}1001 \\ \mathit{1}1000 \\ \mathit{1}0100 \\ \mathit{10100} \end{array}$$

where the digits in *italics* are redundancy bits, one for each row and column of the original 4×4 matrix and also one (in the lower left corner) for the other redundancy bits. Suppose, now, that this set of 25 bits is transmitted from one computer register to another and arrives in the condition indicated by the right side of

$$1984 \quad = \begin{array}{r} \mathit{1}0001 \\ \mathit{0}1001 \\ \rightarrow\mathit{1}100\underline{1} \\ \mathit{1}0100 \\ \mathit{10100} \\ \uparrow \end{array}$$

where the single error has been underlined. The redundancy bits had been chosen for *even* parity and a check of each column and row will reveal that those indicated by arrows fail to have the desired even parity. This leaves little doubt that the error lies in the bit already underlined. By changing the erroneous bit to its opposite value, the desired correction is effected.

R. W. Hamming indicated in his 1950 article that he had found no references to error-correcting codes that predated Marcel Golay's 1949 article.[15] How quickly the theory, if not the actual use, of error-correcting codes mushroomed may be seen by the fact that W. Wesley Peterson published a 270-page book, *Error-Correcting Codes*, in 1961.

Binary Versus Decimal

If most of the space of this chapter has so far been devoted to binary strings for decimal digits, i.e., for *decimal* codes used in *decimal* computers, it is because the use of true binary, in contrast, brought no particular surprises or new developments during the computer age. That true binary could be seen as binary coded octal or hexadecimal (or base 2^n for any n) had already been pointed out in the Burks report and the insight itself goes back at least to Legendre. That this insight in turn makes conversion from binary to octal or hexadecimal a *trivial* matter was also known long before the computer age began. The Burks report also pointed out that conversion from decimal to binary or vice versa, while not trivial, was nevertheless "completely known."

It is not the purpose of this section to discuss the relative merits of binary versus decimal computers, but merely to point out some of the outward manifestations of their inward differences.

Such an outward manifestation can be seen in Table 4. The particular decimal computer used (an IBM 7074) limits mantissas to 8 decimal digits. The particular binary computer used (an IBM 360) similarly limits mantissas to 24 binary digits—at least for ordinary use.

For x = 3, for example, the decimal computer used 0.3333 3333 (instead of the true decimal equivalent of 1/3) and then, upon multiplication by 3, got the result of 0. 9999 9999 that has been recorded in Table 4. The binary equivalent of 1/3 is also infinitely long, namely,

$$1/3 = (0.0101010\overline{01}\ldots)_2.$$

TABLE 4

Comparison of Output from a Decimal and a Binary Computer, Each Having Been Programmed to Compute X(1/X) for Certain Integral Values of X

X	DECIMAL OUTPUT FOR X(1/X)	
	From decimal computer	From binary computer
2	1.0000 0000	1.0000 0000
3	0.9999 9999	0.9999 9994
4	1.0000 0000	1.0000 0000
5	1.0000 0000	0.9999 9994
6	0.9999 9996	0.9999 9976
7	0.9999 9998	0.9999 9994
8	1.0000 0000	1.0000 0000
9	0.9999 9999	0.9999 9994
10	1.0000 0000	0.9999 9964
11	0.9999 9999	0.9999 9970
12	0.9999 9996	0.9999 9976
13	0.9999 9991	0.9999 9994
20	1.0000 0000	0.9999 9994
30	0.9999 9990	0.9999 9994
40	1.0000 0000	0.9999 9994
50	1.0000 0000	0.9999 9994
60	0.9999 9996	0.9999 9994
64	1.0000 0000	1.0000 0000
65	0.9999 9965	0.9999 9994

The binary computer used a truncated version of this, namely,

$$0.0101\ 0101\ 0101\ 0101\ 0101\ 0101,$$

which reads 0.555 555 in hexadecimal. Multiplying this by 3 gives hexadecimal 0.FFF FFF which equals $(16^6 - 1)/(16^6)$, which is approximately equal to decimal 0.9999 9994, as shown in Table 4.

For x = 10, the decimal equivalent of 1/x happens to terminate within the length of the register of the decimal computer. This introduces no truncation error and accounts for the correct entry of 1.000 000. The binary equivalent of 1/10, however, is infinitely long, hence a truncation error is introduced when the binary computer uses only the first 24 digits of an infinitely long mantissa.

It is precisely the binary computer's inability to represent 1/10 exactly that led Daniel McCracken (see page 2) to warn prospective FORTRAN programmers that they would be disappointed if they counted on 1/10 being represented exactly, and 2000 times 1/10 to come to 200 exactly—this being their criterion number for terminating a sequence of operations in a particular sample program—at least if such a program were being run on a binary computer.

The *Reference Day Book* of 1966, published late in 1965, listed 139 currently available general purpose digital computers. Of those listed 46 were decimal, 72 binary, 4 octal, 15 could do their arithmetic in either binary or decimal, and 2 were declared biquinary. Since the last 2 should probably be counted among the decimal computers and the octal among the binary, the final counts are 76 binary, 48 decimal, and 15 binary and decimal.

Alphanumeric Codes

Since input and output for large-scale electronic computers may involve characters other than the ten decimal digits, there is need to have codes for the letters of the alphabet and other symbols, such as the comma and the period. The 26 let-

TABLE 5

Selected Alphanumeric Codes

SYMBOL	C15	C16	C17	C18
A	0000 00	00000	100000	1100 0001
B	0001 00	00001	110000	1100 0010
C	0010 00	00010	100001	1100 0011
D	0011 00	00011	100011	1100 0100
E	0100 00	00100	100010	1100 0101
F	0101 00	00101	110001	1100 0110
G	0110 00	00110	110011	1100 0111
H	0111 00	00111	110010	1100 1000
I	1000 00	01000	010001	1100 1001
J	1001 00		010011	1101 0001
K	1010 00	01001	101000	1101 0010
L	1011 00	01010	111000	1101 0011
M	1100 00	01011	101001	1101 0100
N	1001 00	01100	101011	1101 0101
0	1110 00	01101	101010	1101 0110
P	1111 00	01110	111001	1101 0111
Q	0000 10	01111	111011	1101 1000
R	0001 10	10000	111010	1101 1001
S	0010 10	10001	011001	1110 0010
T	0011 10	10010	011011	1110 0011
U	0100 10		101100	1110 0100
V	0101 10	10011	111100	1110 0101
W	0110 10	10100	010111	1110 0110
X	0111 10	10101	101101	1110 0111
Y	1000 10	10110	101111	1110 1000
Z	1001 10	10111	101110	1110 1001
+	1010 10			0100 1110
&	1011 10			0101 0000
%	1100 10			0110 1100
#	1101 10			0111 1011
’	1110 10			0111 1101
→	1111 10			0101 1111
0	0000 11			1111 0000
1	0001 11			1111 0001
2	0010 11			1111 0010
3	0011 11			1111 0011
4	0100 11			1111 0100
5	0101 11			1111 0101
6	0110 11			1111 0110
7	0111 11			1111 0111
8	1000 11			1111 1000
9	1001 11			1111 1001
.	1010 11			0100 1001
¢	1011 11			0101 1011
*	1100 11			0101 1100
–	1101 11			0110 0000
/	1110 11			0110 0001
′	1111 11			0110 1011

ters of the alphabet and the 10 decimal digits together obviously require more than 32 binary strings and at least a 6-bit code. Several such codes are shown in Table 5.

Code C15 is the "six bit printer synchronizer code" used in the UNIVAC 9200/9300 series computers. Strictly speaking, C16 and C17 are not alphanumeric codes, since no assignment is shown for the ten decimal digits. Code C16, with a marked similarity to C15, was first devised about 1579, by Sir Francis Bacon, as already mentioned in Chapter 1.

Code C17 is a possible interpretation of the Braille code widely used by the blind, and substantially the same as designed by Louis Braille in 1829. This code assigns 40 of the 64 possible subsets of six raised dots (2 columns of 3) to the letters of the alphabet and some commonly used short words and syllables. For example:

```
    **        *           **            **
C =       O =  *      Y =  *      for = **
              *           **            **
```

Code C17 results from the interpretation of a raised dot as "1" and a potential dot as "0," according to the order

```
1 6
2 5
3 4.
```

Code C18 is used in the IBM 360 series computers. It should be noted that this is a binary computer, i.e., it uses true binary in its arithmetic unit. But since input and output is decimal, a decimal code (shown as part of the alphanumeric code) is also needed.

Peano's alphanumeric code of 1899 has already been shown in Chapter 5.

Reflected or Gray Codes

A reflected decimal code has already been mentioned, namely C5 of Table 2. In its purest form, however, a reflected code involves a permutation of *all* 2^n n-bit strings and assign-

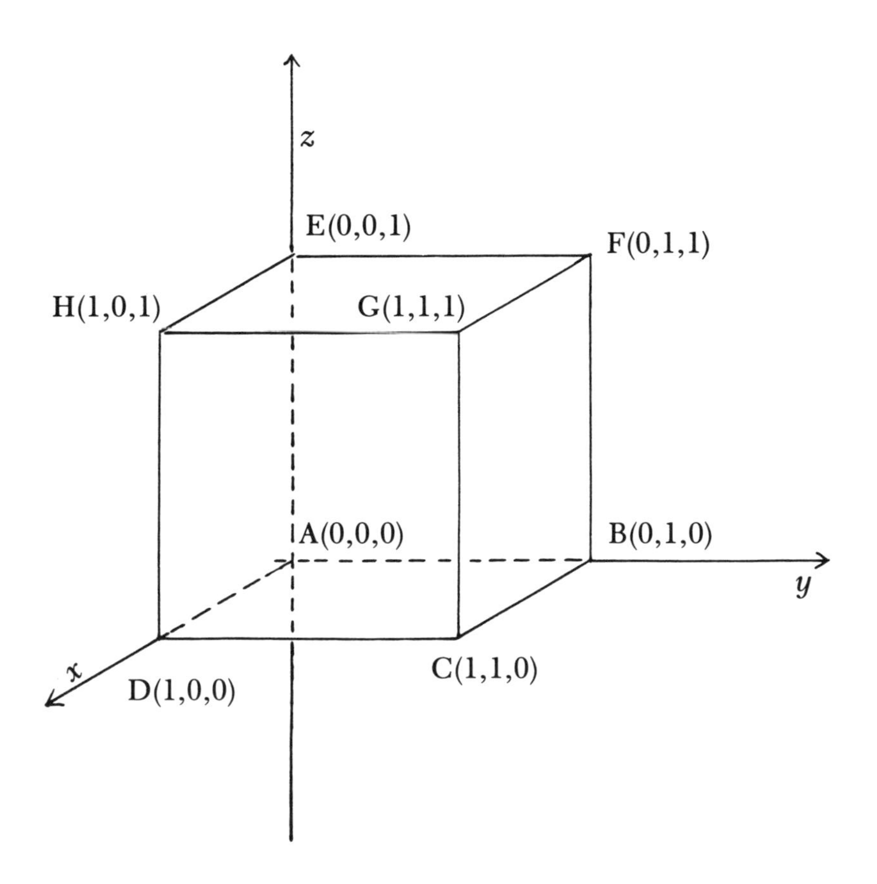

FIGURE 12. Geometric Model for 3-bit Reflected Codes.

ment of them to the integers 0, 1, ..., $(2^n - 1)$. The permutation must show any two successive strings to be identical except for a single one of their n bits. Moreover, the first and last string should again show this minimum difference. Each of the following permutations of the eight 3-bit strings assigned to the 8 octal digits constitutes a reflected code.

	C19	C20	C21
0	000	000	000
1	001	001	010
2	011	011	011
3	010	111	111
4	110	101	110
5	111	100	100
6	101	110	101
7	100	010	001

If an n-bit string is viewed as a n-tuple, and hence also as a point in n-space, and if the distance between two such n-space points is defined as the usual generalization of the comparable formula in 3-space, then the essential property of the permutations involved in a reflected code can be stated as follows: Any two successive strings (the first and last also being considered successive) have a distance of exactly 1. The maximum distance between two n-bit strings is $\sqrt{n}$—the distance between 0...0 and 1...1. Figure 12 shows such an interpretation for 3-space. The question as to how many 3-bit reflected codes are possible can be formulated as: How many paths exist along the edges of a cube from point A(0,0,0) that go through each vertex once and return to point A?

E. N. Gilbert addressed himself to this question for n-bit *reflected codes in general.*[16] For $n \leq 4$ he exhibited all such codes. He found that the number of such codes increases enormously with n, but he was unable to come up with a precise formula.

Without knowing the exact number possible of n-bit re-

TABLE 6

N-bit Reflected Codes for N = 1 *to* N = 5 *that Suggest a Method of Generating Longer Reflected Codes*

DECIMAL	C22	C23	C24	C25	C26
0	0	00	000	0000	00000
1	1	01	001	0001	00001
2		11	011	0011	00011
3		10	010	0010	00010
4			110	0110	00110
5			111	0111	00111
6			101	0101	00101
7			100	0100	00100
8				1100	01100
9				1101	01101
10				1111	01111
11				1110	01110
12				1010	01010
13				1011	01011
14				1001	01001
15				1000	01000
16					11000
17					11001
18					11011
19					11010
20					11110
21					11111
22					11101
23					11100
24					10100
25					10101
26					10111
27					10110
28					10010
29					10011
30					10001
31					10000

flected codes, it is nevertheless easy to generate one in almost random fashion by starting with the 0...0 string and making a single change at a time.

A more systematic way of generating an n-bit reflected code is suggested in Table 6. Here C22 is imbedded in C23, which in turn is imbedded in C24 and so on.

As early as 1954, F. A. Foss reported that up to that time reflected codes had found use only in analog to digital conversions.[17] Here, a number represented by the amount a shaft has turned is converted into digital form. For this application it is particularly important that the first and last strings are also only a single change apart, since they will occupy adjacent positions on the shaft. Foss reported that reflected codes are also useful in digital control systems and urged their use for this purpose. At that time such codes were already known as Gray codes because of a patent that had been issued to F. Gray a year earlier.

NOTES TO CHAPTER VII

[1]R. K. Richards, *Electronic Digital Systems*, (New York: Wiley, 1966), p. 5.

[2]H. H. Goldstine and Adele Goldstine, "The Electronic Numerical Integrator and Computer (ENIAC)," *Mathematical Tables and Other Aids to Computation* 2 (July 1946):97–110.

[3]Arthur W. Burks, Herman H. Goldstine, and John von Neumann, *Preliminary Discussion of the Logical Design of an Electronic Computing Instrument*, (Princeton, New Jersey: Institute for Advanced Study, 1947), p. 7.

[4]Richards, *Electronic Digital Systems*, p. 8.

[5]Jules A. Larrivee, "A History of Computers, II," *The Mathematics Teacher* 51 (November 1958):541–44.

[6]E. B. Craft, L. F. Morehouse, and H. P. Charlesworth, "Machine Switching Telephone System for Large Metropolitan Areas," *Bell System Technical Journal* 2 (April 1923):53–89.

[7]See B. V. Bowden, *Faster Than Thought*, (London: Sir Isaac Pitman and Sons, Ltd., 1953), p. 33; and C. E. Wynn-Williams, "A Thyatron 'Scale of Two' Automatic Counter," *Proceedings of the Royal Society* (*London*) 136 (1932):312–324.

[8]Richards, *Electronic Digital Systems*, pp. 3–4.

[9]Former Ursinus students assure me that here he became known affectionately as "Mad John" for he did not hesitate to put on roller skates and skate across a demonstration table for the sake of illustrating principles of inertia.

[10]Edmund C. Berkeley, *Giant Brains, or Machines that Think*, (New York: Wiley, 1949), p. 129.

[11]George R. Stibitz, "U.S. Patent No. 2, 486,809—Biquinary System Calculator," (November 1, 1949), column 1, lines 16–26.

[12]The 1960 edition of Van Nostrand's *The International Dictionary of Applied Mathematics* includes the following entry: "BIQUINARY. A form of number representation combining the radices 2 and 5." This is obviously at variance with Stibitz.

[13]R. K. Richards, *Arithmetic Operations in Digital Computers*, (Princeton, New Jersey: D. Van Nostrand Company, 1955); and Willis Ware, *Digital Computer Technology and Design*, (New York, Wiley, 1963).

[14]Richards, *Arithmetic Operations in Digital Computers*, pp. 178–79.

[15]Marcel J. E. Golay, "Notes on Digital Coding," *Proceedings of the I. R. E. (Institute of Radio Engineers)* 37 (June 1949):657; and R. W. Hamming, "Error Detecting and Error Correcting Codes," *The Bell System Technical Journal*, 26 (April 1950):147–60.

[16]E. N. Gilbert, "Gray Codes and Paths on the n-Cube," *Bell System Technical Journal* 36 (May 1958):815–26.

[17]F. A. Foss, "The Use of a Reflected Code in Digital Systems," *I. R. E. Transactions—Electronic Computers* 3 (December 1954):1–6.

·VIII·

CONTEMPORARY LITERATURE

A WEALTH of articles and sections of books on nondecimal numeration appeared after 1946. The discussion in this chapter is limited to those that go beyond previous results or are of special interest because of their actual or potential influence in the schools.

SOME NOTEWORTHY ARTICLES

R. Bellman and H. N. Shapiro (1948) presented results concerning the sums of digits of numbers in dyadic notation.[1] Included among these was, in substance,

Theorem 8.1: *As x tends to infinity, $A(x)$ tends to*

$$\frac{x\log x}{2\log 2},$$

where $A(x)$ denotes the sum of all dyadic digits of all positive integers from 1 to x inclusive.

L. Mirsky (1949) extended these results somewhat, as did M. P. Drazin and J. Stanley Griffith in 1952.[2]

A further extension of these results was presented by Peh-Hsuin Cheo and Sze-Chien Yien, who proved (1955)[3]

Theorem 8.2: *As x tends to infinity, $A(x)$ tends to*

$$\frac{\beta - 1}{2}\,\frac{x\log x}{\log\beta}$$

where $A(x)$ is now the comparable sum of the β-adic digits for any base β.

Kenneth Rose (1957) presented a good historical view of the

octal system from the days of Swedberg (1700) to the computer age.[4]

C. E. Shannon (1950) indicated the possible use for computers of numeration systems that are standard except for use of negative digits (like the Barlow and Cauchy modifications).[5] Such systems would be especially helpful if the radix used were odd, for then the system would be *symmetric* (the number of negative digits equals the number of positive digits). The electrical circuits involved with such a system could be simplified.

In response to Shannon's article Z. Pawlak and A. Wakulicz (1957) investigated the possible advantages of a negative base β and proved:[6]

Theorem 8.3: *Every real number has an expansion to the base β, where β is an integer less than -1.*

Theorem 8.4: *If a number has a finite expansion, that expansion is unique.*

In response to this last article, W. Balasinski and S. Mrowka (1957) worked out a criterion for evenness and oddness of a number N in such a negative base notation.[7] They also worked out a (-2) base division logarithm.

Apparently without benefit of previous work on nonintegral bases, George Bergman (1957), while still a junior high school student, showed the use of base τ (tau) where

$$\tau = (1 + \sqrt{5})/2 = 1.618\ 022\ 98\ldots\ .$$

He showed, for example, that

$$\begin{aligned} 1/2 &= 0.010010010010010010\overline{010}\ldots\ , \\ 5 &= 1000.1001\ldots\ , \quad \text{and} \\ 1 &= 0.11 = 0.1011 = 0.101011 \\ &= 0.10101011. \end{aligned}$$

Bergman indicated that no rational fraction can be terminating in this base system, since that would mean that it were equal to a sum of integral powers of τ.[8]

In 1960 Donald Knuth introduced the "quater-imaginary" numeration system that has the imaginary number (2i) as its base.[8a] In this system every complex number can be represented using only the four digits 0, 1, 2, and 3—moreover, no plus or minus sign is needed. In the following examples the right-hand side is written in this new notation:

$$i = 10.2$$

$$-i = 0.2$$

$$7.75 - 7.5i = 11210.31$$

H. L. Alder (1962) investigated those nonstandard numeration systems whose place values are not the series

$$1, \beta^1, \beta^2, \beta^3, \beta^4, \ldots,$$

but rather

$$1, f_1, f_2, f_3, f_4, \ldots,$$

where the f_i's are any nondecreasing set of integers.[9]

Jiri Klir (1962) broadened the above by permitting negative and nonintegral numbers among his f_i's. He concentrated, however, on decimal codes for possible use in computers and went beyond the results obtained by G. P. Weeg.[10]

Konrad Fialkowski (1963) went beyond previous results involving base (-2) and weighted decimal codes, having in mind their possible usefulness in computers.[11]

J. L. Brown (1964) followed up Alder's investigation of nonstandard numeration systems.[12]

N. J. Fine (1965) addressed himself to the following conjecture:[13]

Conjecture: *$N(x)$ is asymptotically $x/(13^2)$, where $N(x)$ denotes the number of n less than x such that n is a multiple of* 13 *and also has a digit sum that is a multiple of* 13.

Fine proved the following theorem which includes this conjecture as a special case.

Theorem 8.5: *As x tends to infinity, $\frac{N(x)}{x}$ tends to* $1/(p^2)$, where $N(x)$ *denotes the number of n less than x such that $n \equiv a \pmod p$ and $s_n \equiv c \pmod p$, where s_n denotes the β-adic* *digit sum of n, and where a and c are any two residues mod p and where p is prime and does not divide* $(\beta - 1)$.

Walter Penney (1965) indicated that computers might be designed to use the nonreal, complex base, $\beta=(-1+i)$, where $i^2 = -1$, and that numbers of the type $X + Yi$ could then be represented, where X and Y are either integers or of the form $k/(2^n)$.[14]

G. F. Songster (1963) gave algorithms for negative base arithmetic, particularly for $\beta = -2$, in which, for example,[15]

$$-35/12 = 1101.01\overline{10}\dots\ .$$

L. C. Eggan and C. L. Vanden Eynden (1966) further investigated the use of rational bases β and the β-adic strings of rational numbers.[16] Among their results is

Theorem 8.6: *If* β *is rational, then any number has at most one periodic expansion to the base* β.

Karl Kieswetter (1966) applied base 4 notation toward constructing a simple example of a function which is everywhere continuous and nowhere differentiable.[17]

Edgar Karst (1967) worked out some algorithms that used two different standard numeration systems simultaneously.[18]

Amount and Extent of Coverage of Numeration Systems in College Textbooks for Future Elementary School Teachers

The only content course in Mathematics (as distinct from one in the teaching of mathematics) usually required of future elementary school teachers is likely to use a text like one of the twelve in the following numbered list:

1. Armstrong, James W. *Mathematics for Elementary School Teachers*. New York: Harper & Row, 1968.
2. Bouwsma, Ward D., Clyde G. Corle, and Davis F. Clemson, Jr. *Basic Mathematics for Elementary Teachers*. New York: The Ronald Press, 1967.

3. Byrne, J. Richard. *Modern Elementary Mathematics*. New York: McGraw-Hill, 1966.
4. Copeland, Richard W. *Mathematics and the Elementary Teacher*. Philadelphia: W. B. Saunders, 1966.
5. Crouch, Ralph and others. *Preparatory Mathematics for Elementary Teachers*. New York: Wiley, 1965.
6. Kovach, Ladis D. *Introduction to Modern Elementary Mathematics*. San Francisco: Holden-Day, 1966.
7. Moise, Edwin E. *The Number Systems of Elementary Mathematics*. Reading, Mass.: Addison-Wesley, 1966.
8. Peterson, John A. and Joseph Hashisaki. *Theory of Arithmetic*. New York: Wiley, 1963.
9. Schaaf, William L. *Basic Concepts of Elementary Mathematics*. New York: John Wiley, 1961.
10. Swain, Robert L. *Understanding Arithmetic*. New York: Holt, Rinehart and Winston, 1965.
11. Webber, G. Cuthbert and John A. Brown. *Basic Concepts of Mathematics*. Reading, Mass.: Addison-Wesley, 1963.
12. Webber, G. Cuthbert. *Mathematics for Elementary Teachers*. Reading, Mass.: Addison-Wesley, 1967.

Table 7 presents some indications as to the coverage of nondecimal numeration systems in these texts. For comparison, the number of pages devoted to ancient numeration systems is also given.

This analysis was made because the coverage in such texts is likely to represent an upper bound of the amount and extent nondecimal standard numeration can be covered in elementary schools, irrespective of what appears in elementary textbooks themselves.

All but one of the twelve texts explicitly indicated a connection between the binary system and computers. None of them covered divisibility rules for nondecimal bases. All but three included multiplication and addition tables for at least one nondecimal base. β-adic representation of fractional numbers was taken up by only two of them. Applications to recrea-

TABLE 7

Content Analysis of Twelve College Textbooks for Future Elementary Teachers with Respect to Numeration Systems

BOOK NUMBER	1	2	3	4	5	6	7	8	9	10	11	12
Number of pages in *entire* book	298	320	410	309	487	223	238	269	364	351	302	145
Number of pages on *ancient* systems	0	8	8	10	4	2	5	19	9	28	9	8
Number of pages on *standard nondecimal* systems	13	5	8	5	16	11	30	17	9	15	10	11
Favorite nondecimal base	8	7	3	5	3	5	2	5	2	5	5	5
Other nondecimal bases treated	2,5 12	2,8	2,5 7,12	2,12	9,11 26	8,2 12	5	2,12	3,5 6,12	2 12	2,7 12	2,3 7
Addition table for at least one base	yes	no	no	no	yes	yes	yes	yes	yes	yes	yes	yes
Multiplication table	yes	no	no	no	yes	yes	yes	yes	yes	yes	yes	yes
β-adic fractions	no	no	no	no	yes	no	no	no	no	yes	yes	yes
Game of NIM	no	no	no	no	no	no	no	yes	no	no	no	no
Guess the number cards	no	no	no	no	no	no	no	yes	no	yes	yes	yes

tional mathematics were limited to NIM (one of the books) and guess the number cards (four of the books).

Nondecimal Numeration in the SMSG (School Mathematics Study Group) Seventh Grade Material

Volume I of *Mathematics for the Junior High School* was intended for the seventh grade and has been available and widely used since the Fall of 1961. The foreword states: "It should be thought of as a sample of the kind of improved curriculum that we need and as a source of suggestions for the authors of commercial textbooks."

Its second chapter, "Numeration," devotes 5, 8, and 31 pages to ancient, decimal, and nondecimal systems respectively. Those 31 pages are about 5% of the 623 pages that constitute the SMSG seventh grade material.

Base 7 is singled out for fairly thorough treatment. Addition and multiplication tables are developed and examples of all four arithmetic operations are given. Some of the exercises are designed to alert the student to the fact that the familiar divisibility rules are tied to base 10 and need to be modified for base 7 numerals. Bases 2, 5, 6, 12, and 60 are also given some attention. For base 2, panels of lights and other examples indicate its special suitability for computers.

For changing a number N from base 10 to base β notation, the method of dividing N by the highest power of β that goes into it (and the resulting remainder by the next highest, and so on) is shown. An attempt at justifying this procedure is included. Several other methods are suggested through exercises or "brainbusters." The entire chapter on numeration systems is restricted to whole numbers.

Contemporary References for Teachers

Future secondary teachers may be exposed to nondecimal numeration in a course in the theory of numbers (usually not required), in some survey course on modern mathematics, or perhaps in a course on mathematics education. The future elementary teacher, however, is more certain of exposure. Aside from books already reviewed and still in print or otherwise widely available, the following are good references in English on the topic of nondecimal numeration.

One of the most comprehensive references—treating fractional as well as whole numbers—is *Topics in Mathematics for Elementary School Teachers*, published by the National Council of Teachers of Mathematics in 1964.

Applications to recreational mathematics (NIM, Chinese Ring Puzzle, Tower of Hanoi) are treated in Edward Kasner and James Newman's *Mathematics and the Imagination*.

Very concise discussions of β-adic notation may be found in Olmstead's book and also in Grace Bates and Fred Kiokemeister's—both entitled *The Real Number System*.

Invitation to Mathematics by William Glenn and Donovan Johnson includes a particularly good layman's description of how binary strings can be transmitted through pulsating signals.

Oystein Ore's *Invitation to Number Theory* contains some material on "Which base is best?" that is probably not otherwise available.

Solved and Unsolved Problems in Number Theory by Daniel Shanks gives a fascinating glimpse into the perfection of perfect numbers as revealed by the reciprocals of their divisors when written in binary notation.

Sherman Stein's *Mathematics, The Man-made Universe* contains a chapter on "Memory Wheels" that explores certain permutations of n-bit strings and their applications to information transmission.

The Bible Dates Itself, which was privately published by Arthur Earle in 1974, is a surprising source on nondecimal numeration. Consider that Adam was 130 years old when Seth was born, that Joseph died when 110, and that Solomon was a judge for 20 years. But if you assume, as Earle does, that these numbers were written in base seven, we find that the numbers shrink to 70, 56, and 14 respectively when translated into base

10 notation. Aside from making these numbers more plausible, Earle's assumptions, for which he gives rather persuasive arguments, result in a good alignment of biblical events with otherwise known historical reference points—and hence the title.

For a particularly good, in-depth, and comprehensive treatment of positional number systems see Chapter 4 of Knuth's 1981 *Seminumerical Algorithms.*

NOTES TO CHAPTER VIII

[1]R. Bellman and H. N. Shapiro, "On a Problem in Additive Number Theory," *Annals of Mathematics* 49 (1948):333–40.

[2]L. Mirsky, "A Theorem on Representations of Integers in the Scale of r," *Scripta Mathematica* 15 (1949):11–12; and M. P. Drazin and J. Stanley Griffith, "On the Decimal Representation of Integers," *Proceedings of the Cambridge Philosophical Society*, 48 (1952):555–65.

[3]Peh-Hsuin Cheo and Sze-Chien Yien, "A Problem on the k-adic Representation of Integers," *Acta Mathematica Sinica* 5 (1955):433–38.

[4]Kenneth Rose, "Tradition vs. Octonary Arithmetic," *The New Philosophy* 60 (January 1957):142–147.

[5]C. E. Shannon, "A Symmetrical Notation for Numbers," *American Mathematical Monthly* 57 (1950):90–93.

[6]Z. Pawlak and A. Wakulicz, "Use of Expansions with a Negative Basis in the Arithmometer of a Digital Computer," *Bulletin de l'academie polonaise des sciences* 5 (March 1957):223–36.

[7]W. Balasinski and S. Mrowka, "On Algorithms of Arithmetical Operations," *Bulletin de l'academie polonaise des sciences* 5 (June 1957):803–804.

[8]George Bergman, "A Number System with an Irrational Base," *Mathematics Magazine* 31 (November-December 1957):98–110.

[8a]Donald E. Knuth, "An Imaginary Number System," *Communications of the Association for Computing Machinery* 3 (1960):245–47.

[9]H. L. Alder, "The Number System in More General Scales," *Mathematics Magazine* 35 (May 1962):145–51.

[10]Jiri Klir, "Weight Codes," *Stroja na spracovani informci, sbornik* 8 (1962):155–62; and G. P. Weeg, "Uniqueness of Weighted Code Representations," *I. R. E. Transactions on Electronic Computers* 9 (December 1960): 487–89.

[11]Konrad R. Fialkowski, "Complex Method of Overflow Determination and Investigation in Binary Weight Codes with Radix −2," *Bulletin de l'academie polonaise des sciences* 11 (September 1963):61–68.

[12]J. L. Brown, Jr., "Generalized Bases for the Integers," *American Mathematical Monthly* 71 (November 1964):973–80.

[13]N. J. Fine, "Distribution of the Sum of the Digits (mod p)," *Bulletin of the American Mathematical Society* 71 (July 1965):651–52.

[14]Walter Penney, "A 'Binary' System for Complex Numbers," *Journal of the Association for Computing Machinery* 12 (April 1965):247–48.

[15]G. F. Songster, "Negative-Base Number-Representation Systems," *IEEE Transactions on Electronic Computers* (EC-12)(June 1963):274–77.

[16]L. C. Eggan and C. L. Vanden Eynden, "'Decimal' Expansions to Nonintegral Bases," *American Mathematical Monthly* 73 (June-July 1966): 576–82.

[17]Karl Kiesswetter, "Ein einfaches Beispiel für eine Funktion welche überall stetig und nicht differenzierbar ist," *Semesterberichte* 13 (1966):216–21.

[18]Edgar Karst, "Algorithms that Use Two Number Systems Simultaneously," *Mathematics Magazine* 40 (March 1967):91–97.

·IX·

SUMMARY

CHRONOLOGY OF IMPORTANT EVENTS IN THE DEVELOPMENT OF NUMERATION SYSTEMS SINCE 1500 A.D.

1500s. The Hindu-Arabic Numeration system took a firmer hold in Western civilization, relegating Roman numerals to such minor uses as numbering chapters of books.

1585. Simon Stevin, Dutch mathematician, published his *De Thiende*, which resulted in decimal fractions becoming a regular part of school arithmetic throughout Europe.

1600 (circa). Thomas Hariot, an English mathematician, was the first person on record to use the binary system. However, this did not become known until 1951, when J. W. Shirley reported on Hariot's unpublished manuscripts. Hariot had been to America to survey the colony of Virginia on behalf of Sir Walter Raleigh. It is possible that he first thought of the binary system during the long ocean voyages to and from America.

1623. Francis Bacon published his *De Augmentis Scientarum*. The first known binary code for the letters of the alphabet is contained therein.

1665. Blaise Pascal published divisibility rules so general as to be applicable to numerals in any base β. He showed specific examples only for bases 10 and 12.

1670. Juan Caramuel y Lobkowitz, a learned bishop in Rome,

was the first to *publish* specific examples of base 2 representations of numbers. He also treated bases 3, 4, 5, 6, 7, 8, 9, 12, and 60. His publication went unnoticed allowing Leibniz to be hailed as the discoverer of the binary system 33 years later.

1672. Erhard Weigel, professor of mathematics at Jena, published on base 4 numerals. He was one of Leibniz's teachers.

1687. Joshua Jordaine's *Duodecimal Arithmetick* appeared in London. He advocated base 12 numerals—at least for fractional numbers—because they are more practical when dealing with fractions of a foot or other measures divided into 12 parts.

1697. Gottfried Leibniz (1646–1716) wrote a letter to the Duke of Brunswick about the binary system. Unaware both of Lobkowitz's published work and of Hariot's unpublished work, he discovered the binary system on his own and had been mentioning it in private correspondence for about a decade. In the letter to the Duke, he suggested that a medallion be struck to commemmorate the discovery of the binary system. The Duke, then 70, ignored the suggestion.

1703. Leibniz's article, "Explication de l'arithmétique binaire," appeared in the official journal of the French Academy of Science along with editorial commentary hailing him as the discoverer of this new arithmetic. The same commentary identified Lagny as a simultaneous discoverer of the binary system.

1708. Emanuel Swedberg, later known as Swedenborg, proposed that the decimal system be replaced for common use by the octal numeration system and that weights and measures be also octalized. Swedberg and the King of Sweden, who strongly supported this matter for his domains, had considered base 64 at first, but had rejected this because it required too many symbols. They agreed that the base of the numeration system ought to be some power of 2, i.e., one of the numbers 2, 4, 8, 16, 32, 64, etc. They settled on 8. However, the king was killed by a cannon ball soon thereafter and with him was buried the proposal to octalize numeration in Sweden.

1798. Adrien-Marie Legendre apparently had the insight that a binary string is at once a binary-coded base 64 string, or a binary-coded base 2^n string for any positive integer n. With the computer age began wide use of this insight, which makes conversion from binary to octal (or hexadecimal) a trivial matter.

1799. The International Metric System of weights and measures took final form and started its path toward international adoption. Since it is a decimalized system, it dovetailed neatly with the decimal numeration system. This made it less likely than ever that a proposal to replace base 10 for common use (such as Swedberg's) would ever find acceptance.

1811. Peter Barlow published his *An Elementary Investigation of the Theory of Numbers* in London. He referred to numeration systems as "scales of notation," gave their divisibility rules, and showed that the nondecimal "scales" were useful in some special situations.

1853. Augustus DeMorgan's *The Elements of Arithmetic* appeared, and contained sections on nondecimal numeration systems. He assumed base 10 was here to stay, but he felt every student should have some experience with nondecimal bases for the greater insight this would give into the common base 10.

1857. Sir Isaac Pitman spoke up strongly against proposed British adoption of the metric system of weights and measures. He recommended instead the more radical reform of changing both the numeration system and the system of weights and measures to a duodecimal (base 12) system. The number 12 is to be preferred as the base, he argued, because 12 is divisible by 2, 3, 4, and 6, whereas 10 is divisible only by 2 and 5. Echos of his position are heard to this day in Great Britain and the United States of America from opponents to full adoption of the decimalized metric system.

1862. John William Nystrom proposed base 16 as a replacement for base 10. He offered a system of weights and measures also based on 16. Four years later he advocated base 12. Apparently he was primarily anti-metric system and hence antidecimal.

1897. Great Britain became the last major country to legalize optional use of the metric system in trade. The United States of America had done so in 1866.

1901. C. L. Bouton showed that the binary system was the key to a complete mathematical theory of the game of NIM.

1946. The first large-scale, general purpose electronic digital computer, the ENIAC, was completed in Philadelphia. While the ENIAC itself did not use true binary, the people involved in the project formulated principles of computer design later reported in the Burks-Goldstein-von Neumann Report. Special suitability for computers of the binary system was here brought out strongly, with appropriate influence on the design of subsequent computers.

1950s. The binary system (in pure or modified form) became known widely as the way to represent numbers inside the computers. Convenient methods of "shrinking" the long binary expressions resulted in octal or hexadecimal representation of numbers. For this reason, Swedberg's octal and Nystrom's hexadecimal (base 16) numeration systems became popular for special purposes, if not for common use.

1965. The binary system played a role in the transmission of the first photographs of Mars.

WHICH BASE IS BEST?

Ten was a poor choice, argued Pascal in 1665. He and many later writers looked for *richness in divisors* (and consequent simpler fractions) and indicated their preference for 12 as the base of our numeration system. For the same reason, other writers advocated 6 or a multiple of 6. Gelin contended that, relative to its size, 8 was even richer in divisors.

On the assumption that the base of the numeration system ought to *dovetail* with the system of weights and measures, the French Metric Commission of the 1790s considered recommending the adoption of base 12 for common arithmetic with the system of weights and measures similarly duodecimalized. Lagrange, who pretended to see no advantage in richness in divisors, argued for getting such dovetailing by decimalizing weights and measures and leaving the arithmetic intact. His view prevailed.

Stein (1826) and Hankel (1874) applied the criterion of the *least number of concepts* or names needed to express all numbers up to some limit L. For L = 1 000 000 base 4 comes out best.

According to Thiele (1889), taking into account *individual differences* among pupils would be particularly easy with base 4. Only 9 significant addition and multiplication facts would have to be learned by the slower students, whereas the faster ones could learn additional facts.

Swedberg (1708) and many later writers favored some power of 2 as a base. This would permit *several halvings* without encountering fractional numbers. His king, Charles XII, went as far as 64 for this purpose. The general tendency was to stay closer to 10 by choosing 8 or 16—as later recommended for use in computers. As early as 1887 Berdellé pointed out that base 8 would at once provide the benefits of base 2, since conversion between these two bases is a trivial procedure.

Lagny (1703) saw in the binary system a *computational tool* to help him solve navigational problems. Leibniz (1703), in contrast, dreamed of the new arithmetic as a *key to theoretical advances*. In the long run the facts sustained Lagny's view.

Applications to electronic computers and other information machines overshadowed the minor theoretical uses.

It has been argued since DeMorgan's time (1853) that a pupil would gain greater insight into common base 10 arithmetic if he had some acquaintance with nondecimal bases. However, such *pedagogical use* does not point to a particular nondecimal base. In practice the relevant textbooks of the 1960s have often singled out either base 5 or base 7.

APPENDICES

TABLE A

Binary Representation of Numbers Compared to more Common Ones

nglish ame	Tally Marks	Roman Numerals	Decimals	Binary Numerals
ero			0	0
ne	/	I	1	1
wo	//	II	2	10
hree	///	III	3	11
our	////	IV	4	100
ive	𝍸	V	5	101
ix	𝍸 /	VI	6	110
even	𝍸 //	VII	7	111
ight	𝍸 ///	VIII	8	1000
ine	𝍸 ////	IX	9	1001
en	𝍸 𝍸	X	10	1010
leven	𝍸 𝍸 /	XI	11	1011
welve	𝍸 𝍸 //	XII	12	1100
hirteen	𝍸 𝍸 ///	XIII	13	1101
ourteen	𝍸 𝍸 ////	XIV	14	1110
ifteen	𝍸 𝍸 𝍸	XV	15	1111
ixteen	𝍸 𝍸 𝍸 /	XVI	16	10000
eventeen	𝍸 𝍸 𝍸 //	XVII	17	10001
ighteen	𝍸 𝍸 𝍸 ///	XVIII	18	10010
ineteen	𝍸 𝍸 𝍸 ////	XIX	19	10011
wenty	𝍸 𝍸 𝍸 𝍸	XX	20	10100

TABLE B

Comparison of Bases Two to Sixteen

English Name	Base 2	Base 3	Base 4	Base 5	Base 6	Base 7	Base 8	Base 9	Base 10	Base 11	Base 12	Base 13	Base 14	Base 15	Base 16
one	1	1	1	1	1	1	1	1	1	1	1	1	1	1	1
two	10	2	2	2	2	2	2	2	2	2	2	2	2	2	2
three	11	10	3	3	3	3	3	3	3	3	3	3	3	3	3
four	100	11	10	4	4	4	4	4	4	4	4	4	4	4	4
five	101	12	11	10	5	5	5	5	5	5	5	5	5	5	5
six	110	20	12	11	10	6	6	6	6	6	6	6	6	6	6
seven	1000	21	13	12	11	10	7	7	7	7	7	7	7	7	7
eight	1000	22	20	13	12	11	10	8	8	8	8	8	8	8	8
nine	1001	100	21	14	13	12	11	10	9	9	9	9	9	9	9
ten	1010	101	22	20	14	13	12	11	10	A	A	A	A	A	A
eleven	1011	102	23	21	15	14	13	12	11	10	B	B	B	B	B
twelve	1100	110	30	22	20	15	14	13	12	11	10	C	C	C	C
thirteen	1101	111	31	23	21	16	15	14	13	12	11	10	D	D	D
fourteen	1110	112	32	24	22	20	16	15	14	13	12	11	10	E	E
fifteen	1111	120	33	30	23	21	17	16	15	14	13	12	11	10	F
sixteen	10000	121	100	31	24	22	20	17	16	15	14	13	12	11	10
seventeen	10001	122	101	32	25	23	21	18	17	16	15	14	13	12	11
eighteen	10010	200	102	33	30	24	22	20	18	17	16	15	14	13	12
nineteen	10011	201	103	34	31	25	23	21	19	18	17	16	15	14	13
twenty	10100	202	110	40	32	26	24	22	20	19	18	17	16	15	14
twenty-one	10110	210	111	41	33	30	25	23	21	1A	19	18	17	16	15
twenty-two	10110	211	112	42	34	31	26	24	22	20	1A	19	18	17	16
twenty-three	10111	212	113	43	35	32	27	25	23	21	1B	1A	19	18	17
twenty-four	11000	220	120	44	40	33	30	26	24	22	20	1B	1A	19	18
twenty-five	11001	221	121	100	41	34	31	27	25	23	21	1C	1B	1A	19
twenty-six	11010	222	122	101	42	35	32	38	26	24	22	20	1C	1B	1A
twenty-seven	11011	1000	123	102	43	36	33	30	27	25	23	21	1D	1C	1B
twenty-eight	11100	1001	130	103	44	40	34	31	28	26	24	22	20	1D	1C
twenty-nine	11101	1002	131	104	45	41	35	32	29	27	25	23	21	1E	1D
thirty	11110	1010	132	110	50	42	36	33	30	28	26	24	22	20	1E
thirty-one	11111	1011	133	111	51	43	37	34	31	29	27	25	23	21	1F
thirty-two	100000	1012	200	112	52	44	40	35	32	2A	28	26	24	22	20

TABLE C

Binary Representation Contrasted with BCD for Selected Numbers (BCD = BINARY-CODED DECIMALS)

Decimal	BCD	Binary
0	0000	0
1	0001	1
2	0010	10
3	0011	11
4	0100	100
5	0101	101
6	0110	110
7	0111	111
8	1000	1000
9	1001	1001
10	0001 0000	1010
11	0001 0001	1011
12	0001 0010	1100
13	0001 0011	1101
14	0001 0100	1110
15	0001 0101	1111
16	0001 0110	10000
17	0001 0111	10001
18	0001 1000	10010
19	0001 1001	10011
20	0010 0000	10100
21	0010 0001	10101
22	0010 0010	10110
...	...	...
32	0011 0010	100000
65	0110 0101	1000001
96	1001 0110	1100000
99	1001 1001	1100011
100	0001 0000 0000	1100100
128	0001 0010 1000	10000000
256	0010 0101 0110	100000000
512	0101 0001 0010	1000000000
1024	0001 0000 0010 0100	10000000000
1984	0001 1001 1000 0100	11111000000
2048	0010 0000 0100 1000	100000000000
4096	0100 0000 1001 0110	1000000000000
8192	1000 0001 1001 0010	10000000000000
16384	0001 0110 0011 1000 0100	100000000000000
32767	0011 0010 0111 0110 0111	111111111111111
32768	0011 0010 0111 0110 1000	1000000000000000
32769	0011 0010 0111 0110 1001	1000000000000001

TABLE D

Comparison of Hexadecimal and Decimal Numerals

hex	dec	hex	dec	hex	dec	hex	dec	hex	dec	hex	dec
1	1	10	16	100	256	1 000	4 096	10 000	65 536	100 000	1 048 576
2	2	20	32	200	512	2 000	8 192	20 000	131 072	200 000	2 097 152
3	3	30	48	300	768	3 000	12 288	30 000	196 608	300 000	3 145 728
4	4	40	64	400	1 024	4 000	16 384	40 000	262 144	400 000	4 194 304
5	5	50	80	500	1 280	5 000	20 480	50 000	327 680	500 000	5 242 880
6	6	60	96	600	1 536	6 000	24 576	60 000	393 216	600 000	6 291 456
7	7	70	112	700	1 792	7 000	28 672	70 000	458 752	700 000	7 340 032
8	8	80	128	800	2 048	8 000	32 768	80 000	524 288	800 000	8 388 608
9	9	90	144	900	2 304	9 000	36 864	90 000	589 824	800 000	9 437 184
A	10	A0	160	A00	2 560	A 000	40 960	A0 000	655 360	A00 000	10 485 760
B	11	B0	176	B00	2 816	B 000	45 056	B0 000	720 896	B00 000	11 534 336
C	12	C0	192	C00	3 072	C 000	49 152	C0 000	786 432	C00 000	12 582 912
D	13	D0	208	D00	3 328	D 000	53 248	D0 000	851 968	D00 000	13 631 488
E	14	E0	224	E00	3 584	E 000	57 344	E0 000	917 504	E00 000	14 680 064
F	15	F0	240	F00	3 840	F 000	61 440	F0 000	983 040	F00 000	15 728 640

Conversion from hexadecimal to decimal or vice versa can be done as follows:

Example 1: $A9F_{hex} = \underline{?}_{dec}$

A9F does not appear in table, but A00, 90, and F *do*:

Hence:

$A00_{hex} = 2560$
$90_{hex} = 144$
$F_{hex} = 15$

$A9F_{hex} = 2719_{dec}$

Example 2: $1995_{dec} = \underline{?}_{hex}$

Since 1995 (dec) does not appear in table, find next lower entry. It is 1792. The difference (1995 − 1792) = 203 does not appear either. Next lower entry is 192, etc.

Hence:

$1792 = 700_{hex}$
$192 = C0_{hex}$
$11 = B_{hex}$

$1995 = 7CB_{hex}$

TABLE E

Selected Numbers Represented in Various Codes

Excess 3 Code	BCD or 8-4-2-1 Code	Dec	Oct	Binary Grouped in 3-bit Words	Binary Grouped in 4-bit Words	Hex
0011 0011 0011	0000 0000 0000	000	0000	000 000 000 000	0000 0000 0000	000
0011 0011 0100	0000 0000 0001	001	0001	000 000 000 001	0000 0000 0001	001
0011 0011 1100	0000 0000 1001	009	0011	000 000 001 001	0000 0000 1001	009
0011 0100 0011	0000 0001 0000	010	0012	000 000 001 010	0000 0000 1010	00A
0011 0100 1000	0000 0001 0101	015	0017	000 000 001 111	0000 0000 1111	00F
0011 0100 1001	0000 0001 0110	016	0020	000 000 010 000	0000 0001 0000	010
0011 0100 1010	0000 0001 0111	017	0021	000 000 010 001	0000 0001 0001	011
0011 1001 0111	0000 0110 0100	064	0100	000 001 000 000	0000 0100 0000	040
0011 1100 1100	0000 1001 1001	099	0143	000 001 100 011	0000 0110 0011	063
0100 0011 0011	0001 0000 0000	100	0144	000 001 100 100	0000 0110 0100	064
0100 0101 1010	0001 0010 0111	127	0177	000 001 111 111	0000 0111 1111	07F
0100 0101 1011	0001 0010 1000	128	0200	000 010 000 000	0000 1000 0000	080
0100 0101 1100	0001 0010 1001	129	0201	000 010 000 001	0000 1000 0001	081
0101 1000 1000	0010 0101 0101	255	0377	000 011 111 111	0000 1111 1111	OFF
0101 1000 1001	0010 0101 0110	256	0400	000 100 000 000	0001 0000 0000	100
1000 0100 0100	0101 0001 0001	511	0777	000 111 111 111	0001 1111 1111	1FF
1000 0100 0101	0101 0001 0010	512	1000	001 000 000 000	0010 0000 0000	200
1010 1001 1010	0111 0110 0111	767	1377	001 011 111 111	0010 1111 1111	2FF
1010 1001 1011	0111 0110 1000	768	1400	001 100 000 000	0011 0000 0000	300
1100 1100 1100	1001 1001 1001	999	1747	001 111 100 111	0011 1110 0111	3E7

TABLE F

Decimal and Binary Equivalents of Fractions of the Type N/64 from n = 0 to n = 63

n/64	Decimal	Binary	n/64	Decimal	Binary
0/64	.000000	.000000	32/64	.500000	.100000
1/64	.015625	.000001	33/64	.515625	.100001
2/64	.031250	.000010	34/64	.531250	.100010
3/64	.046875	.000011	35/64	.546875	.100011
4/64	.062500	.000100	36/64	.562600	.100100
5/64	.078125	.000101	37/64	.578125	.100101
6/64	.093750	.000110	38/64	.593750	.100110
7/64	.109375	.000111	39/64	.609375	.100111
8/64	.125000	.001000	40/64	.625000	.101000
9/64	.140625	.001001	41/64	.640625	.101001
10/64	.156250	.001010	42/64	.656250	.101010
11/64	.171875	.001011	43/64	.671875	.101011
12/64	.187500	.001100	44/64	.687500	.101100
13/64	.203125	.001101	45/64	.703125	.101101
14/64	.218750	.001110	46/64	.718750	.101110
15/64	.234375	.001111	47/64	.734375	.101111
16/64	.250000	.010000	48/64	.750000	.110000
17/64	.265625	.010001	49/64	.765625	.110001
18/64	.281250	.010010	50/64	.781250	.110010
19/64	.296875	.010011	51/64	.796875	.110011
20/64	.312500	.010100	52/64	.812500	.110100
21/64	.328125	.010101	53/64	.828125	.110101
22/64	.343750	.010110	54/64	.843750	.110110
23/64	.359375	.010111	55/64	.859375	.110111
24/64	.375000	.011000	56/64	.875000	.111000
25/64	.390625	.011001	57/64	.890625	.111001
26/64	.406250	.011010	58/64	.906250	.111010
27/64	.421875	.011011	59/64	.921875	.111011
28/64	.437500	.011100	60/64	.937500	.111100
29/64	.437500	.011100	60/64	.937500	.111100
30/64	.453125	.011101	61/64	.953125	.111101
31/64	.484375	.011111	63/64	.984375	.111111

TABLE G

Binary and Other Equivalents and Approximations of Fractions of the type n/10 from n = 0 to n = 9

Fractions		Equivalents		Approximations	
n/10	Reduced	Decimal	Binary	Truncated Binary	Truncated Hexadecimal
0/10		.0	$.\overline{0}\ldots$	.0000 0000	.00
1/10		.1	$.0\overline{0011}\ldots$	.0001 1001	.19
2/10	1/5	.2	$.\overline{0011}\ldots$	.0011 0011	.33
3/10		.3	$.0\overline{1001}\ldots$	.0100 1100	.4C
4/10	2/5	.4	$.\overline{0110}\ldots$	.0110 0110	.66
5/10	1/2	.5	$.1\overline{0}\ldots$	.1000 0000	.80
6/10	3/5	.6	$.\overline{1001}\ldots$	.1001 1001	.99
7/10		.7	$.1\overline{0110}\ldots$	.1001 0011	.B3
8/10	4/5	.8	$.\overline{1100}\ldots$	.1100 1100	.CC
9/10		.9	$.1\overline{1100}\ldots$	.1110 0110	.E6

TABLE H

Binary and Other Equivalents and Approximations of Unit Fractions from 1/2 to 1/16

	Equivalents		Approximations	
1/n	Decimal	Binary	Truncated Decimal	Truncated Binary
1/2	$.5\overline{0}\ldots$	$.1\overline{0}\ldots$	.500	.1000 0000
1/3	$.\overline{3}\ldots$	$.\overline{01}\ldots$	.333	.0101 0101
1/4	$.25\overline{0}\ldots$	$.01\overline{0}\ldots$	.250	.0100 0000
1/5	$.2\overline{0}\ldots$	$.\overline{0011}\ldots$	.200	.0011 0011
1/6	$.1\overline{6}\ldots$	$.0\overline{01}\ldots$	.166	.0010 1010
1/7	$.\overline{142857}\ldots$	$.\overline{001}\ldots$	.142	.0010 0100
1/8	$.125\overline{0}\ldots$	$.001\overline{0}\ldots$	.125	.0010 0000
1/9	$.\overline{1}\ldots$	$.\overline{000111}\ldots$	.111	.0001 1100
1/10	$.1\overline{0}\ldots$	$.0\overline{0011}\ldots$	.100	.0001 1001
1/11	$.\overline{09}\ldots$	$.\overline{0001011101}\ldots$	.090	.0001 0111
1/12	$.08\overline{3}\ldots$	$.00\overline{01}\ldots$	.083	.0001 0101
1/13	$.\overline{076923}\ldots$	$.\overline{000100111011}\ldots$	.076	.0001 0011
1/14	$.0\overline{714285}\ldots$	$.0\overline{001}\ldots$	.071	.0001 0010
1/15	$.0\overline{6}\ldots$	$.\overline{0001}\ldots$	.066	.0001 0001
1/16	$.0625\overline{0}\ldots$	$.0001\overline{0}\ldots$	.062	.0001 0000

TABLE I

The Number 1/7 Expressed in Bases Two to Sixteen

Base β	β-adic String	Length of Period
two	$0.\overline{001}\ldots$	3
three	$0.\overline{010212}\ldots$	6
four	$0.\overline{021}\ldots$	3
five	$0.\overline{032412}\ldots$	6
six	$0.\overline{05}\ldots$	2
seven	$0.1\overline{0}$	1
eight	$0.\overline{1}\ldots$	1
nine	$0.\overline{125}\ldots$	3
ten	$0.\overline{142857}\ldots$	6
eleven	$0.\overline{163}\ldots$	3
twelve	$0.\overline{186A35}\ldots$	6
thirteen	$0.\overline{1B}\ldots$	2
fourteen	$0.2\overline{0}\ldots$	1
fifteen	$0.\overline{2}\ldots$	1
sixteen	$0.\overline{249}\ldots$	3

FONTENELLE's article "Nouvelle Arithmétique Binaire" deserves to be appended in its entirety and in the original French, so that other researchers may check their interpretation against the one appearing in the present volume (pages 43–45), where it is claimed that Fontenelle has usually been misread in the passages in which he describes Professor Lagny's work with the binary system.

Fontelle's first marginal comment "v. les M. page 85" says "see the *Memoires*, page 85" (where Leibniz's "Explication" appears), his second "page 88" refers to some additional information about Professor Lagny that had appeared in the preceding year's *Histoire*.

Bernard le Bovier de Fontenelle (1657–1757) became "perpetual" Secretary of the Academy of Science in Paris in 1697 and remained in that post for 42 years. In this capacity he was editor of the two chief publications of the Academy, namely, the *Memoires* and the *Histoire*. It was the former that contained Leibniz's "Explication" and the latter the editorial comment by Fontenelle entitled "Nouvelle Arithmétique Binaire." Both appeared in the 1703 volume.

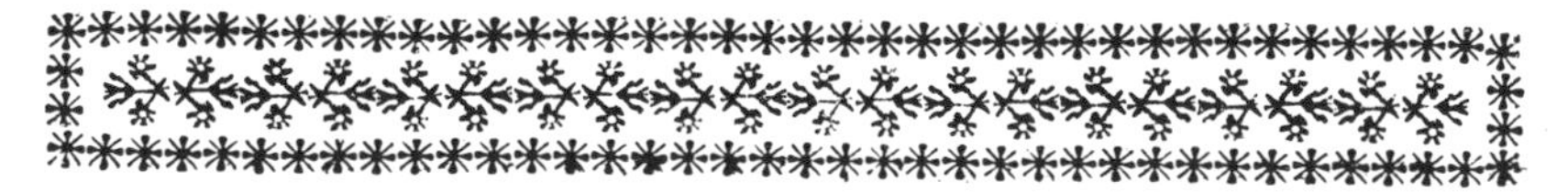

ARITHMETIQUE.

NOUVELLE ARITHMETIQUE BINAIRE.

V. les Mém. p. 85.

LA Science des Nombres eſt ſi naturelle aux Hommes, cultivée depuis tant de ſiécles, & par tant d'Eſprits excellens, pouſſée préſentement à un ſi haut point de perfection, que ce doit être une eſpéce de prodige, qu'une Arithmétique nouvelle, & toute différente de celle que nous pratiquons.

Cependant, à considerer la chose de plus près, le fondement de toute notre Arithmetique étant purement arbitraire, il est permis de prendre une autre fondement, qui nous donnera une autre Arithmétique. On a voulu que la Suite premiére & fondamentale des Nombres allât jusqu'à Dix, & que la Suite infinie des Nombres, fût une suite infinie de Dixaines. Mais il est visible que d'avoir étendu la Suite fondamentale des Nombres jusqu'à Dix, ou de ne l'avoir pas étendue plus loin, c'est une institution qui eût pû être différente. Et même il paroît qu'elle a été faite assez au hasard par les peuples, & que les Mathématiciens n'en ont pas eté consultés; car ils auroient aisément établi quelque chose de plus commode. Par exemple, si l'on eût poussé la Suite des Nombres jusqu'à Douze, on y eût trouvé sans Fraction des Tiers & des Quarts qui ne sont pas dans Dix.

Les Nombres ont deux sortes de propriétés, les unes essentielles, les autres dépendantes d'une institution arbitraire, & de la maniere de les exprimer. Que les Nombres impairs toujours ajoutés de suite, donnent la Suite naturelle des Quarrés, c'est une propriété essentielle à la Suite infinie des Nombres, de quelque maniere qu'on l'exprime. Mais que dans tous les Multiples de 9, les caracteres qui les expriment additionnés ensemble, rendent toujours 9, ou un multiple de 9, moindre que celui qui a été proposé, c'est une propriété qui n'est nullement essentielle au nombre 9, & qu'il n'a que parce qu'il est le penultiéme nombre de la progression décuple qu'il nous a plû de choisir. Si l'on eût pris la progression de Douze, 11 auroit eu la même propriété.

Il est bien commode de pouvoir reconnoître au premier coup d'œil, & sans aucune opération que 25245, par exemple, est un multiple de 9; & si des Mathématiciens avoient établi la progression fondamentale qui devoit regner dans l'Arithmétique, ils auroient, après les avoir toutes examinées, préféré celle qui auroit produit le plus de semblables commodités, soit pour l'usage commun & populaire, soit pour les recherches sçavantes.

M. Leibnitz ayant étudié la plus simple & la plus courte de toutes les progressions possibles, qui est celle qui se termine à Deux, l'a trouvée très-riche & très-abondante en ces sortes de propriétés accidentelles. Il n'y auroit dans toute son Arithmétique que deux caracteres 1 & 0. Le Zéro auroit la puissance de multiplier tout par deux; comme dans l'Arithmétique ordinaire, il multiplie tout par dix. 1 seroit un, 10 deux, 11 trois, 100 quatre, 101 cinq, 110 six, 111 sept, 1000 huit, 1001 neuf, 1010 dix, &c. ce qui est entiérement fondé sur les mêmes principes que les expressions de l'Arithmétique commune.

Il est vrai que celle-ci seroit très-incommode, par la grande quantité de caracteres dont elle auroit besoin, même pour de très-petits nombres. Il lui faut, par exemple, 4 caracteres pour exprimer huit, que nous exprimons par un seul. Aussi M. Leibnitz ne veut-il pas faire passer son Arithmétique dans un usage populaire; il prétend seulement que pour des recherches difficiles, elle aura des avantages que l'autre n'a pas, & qu'elle conduira à des spéculations plus élevées.

Ce fut en 1702 qu'il communiqua à l'Académie cette Arithmétique Binaire, annonçant seulement qu'elle auroit de grands usages pour les Sciences, & ne les découvrant point. Il ne voulut point qu'il en fût parlé dans l'Histoire, jusqu'à ce que cette nouvelle invention pût paroître accompagnée de ses utilités.

Dans la présente année, il se trouva qu'elle en avoit une, à laquelle M. Leibnitz lui-même ne se fût pas attendu. Le P. Bouvet Jésuite, célèbre Missionnaire de la Chine, à qui M. Leibnitz avoit écrit l'idée de son Arithmétique Binaire, lui manda qu'il étoit très-persuadé que c'étoit-là le véritable sens d'une ancienne Enigme Chinoise, laissée il y a plus de 4000 ans par l'Empereur Fohi, Fondateur des Sciences de la Chine, aussi-bien que de l'Empire, entendue apparemment dans son siécle, & plusieurs siécles après lui, mais dont il étoit certain que l'intelligence s'étoit perdue depuis plus de 1000 ans, malgré les recherches & les efforts des plus Sçavans *Lettrés*, qui n'avoient attrappé que des Allégories puériles & chimériques. Cette Enigme consiste dans les différentes combinaisons d'une ligne entiere, & d'une ligne brisée, répétées un certain nombre de fois, soit l'une, soit l'autre. En supposant que la ligne entiere signifie 1, & la brisée 0, on trouve les mêmes expressions de nombres que donne l'Arithmétique Binaire. La conformité des combinaisons des deux lignes de Fohi, & des deux uniques caracteres de l'Arithmétique de M. Leibnitz, frappa le P. Bouvet, & lui fit croire que Fohi & M. Leibnitz avoient eu la même pensée. Si la vérité de cette heureuse rencontre se confirme, quelle gloire pour les Européens, du moins aux yeux des Chinois, de leur avoir donné la Clef de leur ancienne Science! Il est toujours certain qu'en pensant autant que l'on fait présentement, & en tournant d'autant de façons différentes une certaine matiere, & un certain fonds de pensées raisonnables, qui a été donné aux Hommes, il est impossible qu'on ne retrouve à peu près tout ce que les autres siécles auront pensé de meilleur.

Si M. Leibnitz ne s'est pas rencontré sur l'Arithmétique Binaire avec l'Empereur Fohi, du moins M. de Lagni s'est rencontré avec M. Leibnitz sur ce même sujet. M. de Lagni, Professeur en Hydrographie à Rochefort, travaille, comme on l'a déja pû voir dans l'Histoire de 1702,* à perfectionner la Science qu'il professe. Il a entrepris par rapport à la Navigation, une nouvelle Trigonométrie; & en étudiant tout le Systême des Logarithmes, qui ont été inventés principalement pour la Trigonométrie, il y a vû des défauts & des inconveniens, dont il n'a pû trouver le remède qu'en imaginant l'Arithmétique Binaire.

* Pag. 88.

La grande commodité des Logarithmes, eſt de changer les Multiplications & les Diviſions, qui ſont des Opérations longues & difficiles pour les grands nombres, en des Additions ou Souſtractions, qui ſont beaucoup plus ſimples & plus aiſées. Mais M. de Lagni prétend que cet avantage que la Théorie promet ſi magnifiquement, ſe réduit à rien dans la Pratique; qu'au contraire comme les Logarithmes, qui ſont des eſpéces de nombres feints & ſuppoſés, ſont un circuit que l'on prend pour arriver aux Nombres *naturels*, les ſeuls que l'on cherche, il y a toujours plus de chemin à faire, quoique peut-être plus facilement, & toujours un plus long tems à employer, & il en appelle à témoins tous ceux qui ont calculé par cette Méthode. Il avance même que les Logarithmes ſont faux dans les grands nombres; il en donne pour preuve un calcul que Henri Brigs dans ſon Arithmétique Logarithmique *pag.* 27. *& ſuiv.* a donné pour exemple de l'uſage des Logarithmes.

Dans l'Arithmétique Binaire les Multiplications & les Diviſions ſe font néceſſairement par de ſimples Additions & Souſtractions, ſans qu'il faille paſſer par aucun circuit, tel qu'eſt celui des Logarithmes dans l'Arithmétique commune; & par conſéquent tout l'avantage que l'Arithmétique commune ne tire des Logarithmes que par force, eſt eſſentiel à l'Arithmétique Binaire, dont M. de Lagni nomme par cette raiſon les Multiplications & les Diviſions, *Logarithmes naturels.*

Il a mis ſon idée plus au long dans un Ecrit qu'il imprima cette année à Rochefort, & qu'il envoya à l'Académie; mais le peu que nous en avons dit, ſuffira pour mettre ſur les voies, ceux qui voudront approfondir cette nouvelle Arithmétique.

Comme les plus grands Mathématiciens peuvent très-légitimement être jaloux de la gloire de s'être rencontrés avec M. Leibnitz, ſans l'avoir ſuivi, nous devons ici ce témoignage à M. de Lagni, qu'ayant toujours été à Rochefort il ne paroît point avoir eu aucune connoiſſance de ce que M. Leibnitz avoit envoyé à l'Académie ſur le Calcul Binaire.

BIBLIOGRAPHY

Adams, John Quincy. *Report upon Weights and Measures.* Washington, D.C.: United States Department of State, 1817.

Adkins, Julia. "Bibliography on Number Bases," *Arithmetic Teacher*, 6:324, December 1959.

Ahrens, Wilhelm. *Mathematische Unterhaltungen und Spiele.* Leipzig: B. G. Teubner, 1901.

______. *Mathematische Unterhaltungen und Spiele.* Leipzig: B. G. Teubner, 1910.

Alder, H. L. "The Number System in More General Scales," *Mathematics Magazine*, 35:145–151, May, 1962.

Ampère, André-Marie. *Essai sur la philosophie des sciences.* Paris: Bachelier, 1838.

Andrews, F. Emerson. "Decimal-form Fractions to Various Bases," *The Mathematics Teacher*, 32:354–55, December, 1939.

______. "An Excursion in Numbers," *Atlantic Monthly*, 154: 459–66, October, 1934.

______. *New Numbers.* New York: Essential Books, 1944.

______. "Revolving Numbers," *Atlantic Monthly*, 155:208–11, February, 1935.

Archibald, Raymond Clark. "The Binary Scale of Notation, a Russian Peasant Method of Multiplication, the Game of Nim and Cardan's Rings," *The American Mathematical Monthly*, 25:139–142, March, 1918.

Bachet, Claude-Gaspard. *Problèmes plaisant et délectables qui font par les nombres.* 4th ed. Paris: Gauthier-Villars, 1879.

Bachman, Paul. *Niedere Zahlentheorie.* Leipzig: B. G. Teubner, 1902.

Bacon, Francis. *Of the Advancement and Proficiency of Learning.* Trans. Gilbert Wats. Oxford: Young & Forrest, 1640.

Balasinski, W. and S. Mrowka. "On Algorithms of Arithmetical Operations." *Bulletin de l'academie polonaise des sciences*, 5:803–804. June, 1957.

Ball, W. W. Rouse and H. S. M. Coxeter. *Mathematical Recreations and Essays*. New York: Macmillan, 1960.

Barlow, Peter. *A New Mathematical and Philosophical Dictionary*. London: G. and S. Robinson, 1814.

______. "On the Method of Transforming a Number from one Scale of Notation to Another, and its Application to the Rule of Duodecimals," *Journal of Natural Philosophy, Chemistry and the Arts*, 25:181–188, 1810.

______. *Theory of Numbers*. London: J. Johnson, 1811.

Bates, Grace E. and Fred. L. Kiokemeister. *The Real Number System*. Boston: Allyn and Bacon, Inc., 1960.

Begle, E. *Very Short Course in Mathematics for Parents*. Stanford, Cal.: School Mathematics Study Group, 1963.

Beguelin, N. "Application de l'Algorithme exponentiel à la recherche des facteurs des nombres de la forme $2^n + 1$," *Nouveaux Memoires de l'Academie Royale des Sciences et Belles-Lettres*, (Berlin), 1777:239–264, 1777.

Bell, E. T. *The Development of Mathematics*. New York: McGraw-Hill, 1945.

Bellavitis, G. "Sulla risoluzione delle congruenze numeriche e sulle tabole che danno i logaritme (indici) degli interi rispetto ai vari moduli." *Atti, Accademia nationale dei Lincei, Roma, classe di scienze fisiche, matmatiche e naturali*, s. 3, 1:778–800, 1877.

Bellman, R. and H. N. Shapiro, "On a Problem in Additive Number Theory," *Annals of Mathematics*, 49:333–340, 1948.

Berckenkampf, Johann Albert. *Leges numerandi quibus numeratio decadia Leibnizia dyadica*. Lemgovia: n.n., 1747.

Berdellé, Charles. "La numération binaire et la numération octavale," *Association française pour l'avancement des sciences*, 1887:206–209, 1887.

Bergman, George. "A Number System with an Irrational

Base," *Mathematics Magazine*, 31:98–110, November-December, 1957.

Berkeley, Edmund C. *Giant Brains, or Machines that Think*. New York: Wiley, 1949.

______. *The Reference Diary of the Library of Computer and Information Sciences*. New York: The Library of Computer and Information Sciences, 1965.

Bezout, Etienne. *Course de mathématique*, n.p.: n.d.

Blij, F. van der. "Combinatorial Aspects of the Hexagrams in the Chinese Book of Changes," *Scripta Mathematica*, 28:37–49, May, 1967.

Borda, Lagrange, Lavoisier, Tillet, and Condorcet. "Rapport fait a l'Academie des Sciences," *Histoire de l'Academie Royale des Sciences*, 1788:1–6, 1788.

Bouton, C. L. "Nim: A Game with a Complete Mathematical Theory," *Annals of Mathematics*, series 2, 3:35–39, 1901–1902.

Bowden, B. V. *Faster Than Thought*. London: Sir Isaac Pitman and Sons, Ltd., 1953.

Bowden, Joseph. "The Russian Peasant Method of Multiplication," *The Mathematics Teacher*, 5:4–8, 1912.

______. *Special Topics in Theoretical Arithmetic*. Garden City, N.Y.: Joseph Bowden, 1936.

Brander, Georg Friederich. *Arithmetica binaria sive dyadica das ist Die Kunst nur mit zwey Zahlen in allen vorkommenden Fällen sicher und leicht zu rechnen*. Augsburg: The Author, 1775.

Brocard, H. "Question 2960," *Intermédiaire des mathématiciens*, 13:72–73, 1906.

Brockhaus Conversations-Lexicon, 12 vols. Wiesbaden: E. Brockhaus, 1883.

Brooks, Edward. *The Philosophy of Arithmetic*. Philadelphia: Sower, Potts, and Co., 1876.

Brown, Jr., J. L. "Generalized Bases for the Integers," *American Mathematical Monthly*, 71:973–980, November, 1964.

Brunetti, Francisco Saverio, *Aritmetica binomica e diadica*. Rome: Bernabo e Lazzarini, 1746.

Buchanan, Herbert E. and others. *A Brief Course in Advanced Algebra*. Boston: Houghton Mifflin Co., 1937.

Buchholz, W. "Fingers or Fists?" *Communications of the Association for Computing Machinery*, 2:3–11, December, 1959.

Buffon, Georges Louis Leclerc. *Oeuvres complètes*. 12 vols. Paris: Garnier Frères, 1938.

Burks, Arthur W., Herman H. Goldstine, and John von Neumann. *Preliminary Discussion of the Logical Design of an Electronic Computing Instrument*. Princeton, N.J.: Institute for Advanced Study, 1947.

Cajori, Florian. "Leibniz, The Master-builder of Mathematical Notations," *Isis*, 7:412–429, 1925.

Cantor, Georg. *Gesammelte Abhandlungen Mathematischen und Philosophischen Inhalts*. Hildesheim: Georg Olms Verlagsbuchhandlung, 1962.

Cantor, Moritz. *Mathematische Beiträge zum Kulturleben der Völker*. Hall: H. W. Schmidt, 1863.

______. *Vorlesengun über Geschichte der Mathematik*. Leipzig: B. G. Teubner, 1922.

______. "Zahlentheoretische Spielereien," *Zeitschrift für Mathematik und Physik*, 20:134–135 (Hist.-Litt. Abt), 1875.

Caramuel, Juan. *Mathesis biceps*. Campaniae: L. Annison, 1670.

Carus, P. "Chinese Philosophy," *The Monist*, 6:188–249, 1896.

Cauchy, Augustin-Louis. *Oeuvres*. Paris: Gauthier-Villars et Fils, 1885.

Cheo, Peh-Hseuin and Sze-Chien Yien. "A Problem on the k-adic Representation of Integers," *Acta Mathematica Sinica*. 5:433–438, 1955.

Christofferson, H. C. "A New Number System," *School Science and Mathematics*, 24:913–916, December, 1924.

Chrystal, George. *An Elementary Textbook of Algebra*. 2 vols. Edinburgh: A. and C. Black, 1919–20. (other editions earlier and later)

Collignon, Ed. "Note sur l'arithmétique binaire," *Journal de mathématiques elémentaires*, 21:101–106, 126–131, 148–151, 171–174, 1897.

Commission on Mathematics. *The Mathematics of the Seventh and Eighth Grades*. New York: College Entrance Examinations Board, 1957.

Committee on the Undergraduate Program in Mathematics. *Course Guides for the Training of Teachers of Elementary School Mathematics*. Third Draft. Berkeley: Committee on the Undergraduate Program in Mathematics, August 1963.

Conant, Levi Leonard. "Primitive Number Systems," *Smithsonian Institution Annual Report*, 583–594, 1892.

Condon, E. U. "The Nimatron," *American Mathematical Monthly*, 49:330–332, May, 1942.

Cooper, G. H. "Eight-digit System," *Scientific American*, 122:111, January, 1920.

Courant, Richard and Herbert Robbins. *What is Mathematics*. London: Oxford University Press, 1941.

Craft, E. B., L. F. Morehouse, and H. P. Charlesworth. "Machine Switching Telephone System for Large Metropolitan Areas," *Bell System Technical Journal*, 2:53–89, April, 1923.

Cunningham, Allan. "On Binal Fractions." *The Mathematical Gazette*, 4:259–267, 1908.

Dangicourt, Petr. "De periodis columnarum in serie numerorum progressionis arithmeticae dyadice expressorum," *Miscellanea Berolinensia*, 1:336–376, 1710.

Delambre, Jean Baptiste Joseph. "Notice sur la vie et les ouvrages de M. Lagrange," *Memoires de la classe des sciences mathématiques et physiques de l'institute Impérial de France*, 13:XXVI–LXXX, 1812.

DeMorgan, Augustus. *A Budget of Paradoxes*. London: Longmans, Green, and Co. 1872.

______. *The Elements of Arithmetic*. London: Taylor, Walton and Maberly, 1853.

______. *Study and Difficulties of Mathematics.* Chicago: Open Court Publishing Co., 1910.

Dickson, Leonard Eugene. *History of the Theory of Numbers.* 3 vols. New York: G. E. Steckert and Co., 1934.

Drazin, M. P. and J. Stanley Griffith. "On the Decimal Representation of Integers," *Proceedings of the Cambridge Philosophical Society*, 48:555–565, 1952.

Dudley, Underwood. "The First Recreational Mathematics Book," *Journal of Recreational Mathematics*, 3:164–169, July, 1970.

DuPasquier, Louis-Gustave. *Le développement de la notion de nombre.* Paris: Attinger Frères, 1921.

Earle, Arthur. *The Bible Dates Itself.* Southampton, Pennsylvania: Arthur Earle, 1974.

Eggan, L. C. and C. L. Vanden Eynden. "'Decimal' Expansions to Nonintegral Bases," *The American Mathematical Monthly*, 73:576–582, June-July, 1966.

Eisele, Carolyn, ed. *The New Elements of Mathematics by Charles S. Peirce.* Atlantic Highlands, N.J.: Humanities Press, 1976.

Eneström, Gustaf. *Verzeichnis der Schriften Leonhard Eulers.* Leipzig: B. G. Teubner, 1910.

Euler, Leonard. *Opera postuma.* Petropoli: Fuss und Fuss, 1892.

Felkel, Anton. "Verwandlung der Bruchsperioden, nach den Gesetzen verschiedener Zahlensysteme," *Ceska spolecnost Nauk, Prague Abhandlungen*, series 2, 1785:135–174, 1785.

Fialkowski, Konrad R. "Complex Method of Overflow Determination and Investigation in Binary Weight Codes with Radix −2," *Bulletin de l'academie polonaise des sciences*, 11:61–68, September, 1963.

______. "The Properties and Utilization of Semi-systematic Weight Codes," *Bulletin de l'academie polonaise des sciences*, 13:967–974, August, 1965.

______. "The $\bar{2}$'s Complement and Other Semi-Systematic Binary Codes," *IEEE Transactions on Electronic Computers*, 15:603–05, 1966.

Fine, N. J. "Distribution of the Sum of Digits (mod p)," *Bulletin of the American Mathematical Society*, 71:651–652, July, 1965.

Fontenelle, Bernard Le Bovier de. "Nouvelle Arithmétique," *Histoire de l'Academie Royale des Sciences*, 1703:58–63, 1703.

Foss, F. A. "The Use of a Reflected Code in Digital Systems," *IRE Transactions—Electronic Computers*, 3:1–6, December, 1954.

Fraenkel, Abraham Adolf. "Natural Numbers as Cardinals," *Scripta Mathematica*, 6:69–79, June 1939.

Gauss, Carl Friedrich. *Disquistiones arithmeticae*. Trans. Arthur A. Clarke. New Haven, Conn.: Yale University Press, 1966.

Gegenbauer, Leopold. "Über Zahlensysteme," *Sitzungsberichte der Mathematisch-Naturwissenschaftlichen Classe der Kaiserlichen Akademie der Wissenschaften, Wien*, 95:618–627, 1887.

Gelin, E. "Du meilleur système de numération et de poids et mesures," *Mathesis, Recueil Mathématique* (2)6:161–164, 1896.

Gerhardt, C. I. *Leibnizene mathematische Schriften*. 7 vols. Berlin: A. Asher und Comp., 1849.

Gilbert, E. N. "Gray Codes and Paths on the n-Cube," *Bell System Technical Journal*, 36:815–826, May, 1958.

Glaisher, J. W. L. "An Arithmetical Proposition," *Messenger of Mathematics*, 2:188–191, 1873.

Glenn, William H. and Donovan Johnson. *Invitation to Mathematics*. Garden City, New York: Doubleday, 1961.

Golay, Marcel J. E. "Notes on Digital Coding," *Proceedings of the I.R.E. (Institute of Radio Engineers)*, 37:657, June, 1949.

Goldstine, H. H. and Adele Goldstine. "The Electronic Numerical Integrator and Computer (ENIAC)," *Mathe-*

matical Tables and Other Aids to Computation, 2:97–110, July, 1946.

Goldstine, Herman H. *The Computer from Pascal to von Neumann*. Princeton: Princeton University Press, 1972.

Grossman, H. D. and D. Kramer. "A New Match Game," *American Mathematical Monthly*, 52:441–443, October, 1945.

Grünwald, Vittorio. "Intorno all'aritmetica dei sisteme numerici a base negativa," *Giornale di matematiche di Battaglini*, 23:203–221, 367, 1885.

Guhrauer, G. E. *Gottfried Wilhelm Freiherr von Leibniz*. 2 vols. Breslau: Ferdinand Hirt, 1846.

Hamming, R. W. "Error Detecting and Error Correcting Codes," *The Bell System Technical Journal*, 26:147–160, April, 1950.

______. "The Binary System in Modern Times." Paper presented at the Festkolloquium 300 Jahre Dualsystem, on March 21 1979, at the Bavarian Academy of Sciences in Munich.

Hankel, Hermann. *Zur Geschichte der Mathematik in Alterthum und Mittelalter*. Leipzig: B. G. Teubner, 1874.

Hardy, G. H. and E. M. Wright. *An Introduction to the Theory of Numbers*. 3rd ed. Oxford: Clarendon Press, 1954.

Hariot, Thomas. "Mathematical Calculations and Annotations," (Film 128 in the library of the University of Michigan, original manuscript in British Museum).

Harkin, Duncan. *Fundamental Mathematics*. New York: Prentice-Hall, 1941.

Heath, Thomas L. *A Manual of Greek Mathematics*. Oxford: Clarendon Press, 1931.

Hochstetter, Erich et al. *Herrn von Leibniz' Rechnung mit Null und Eins*. Berlin: Siemens Aktiengesellschaft, 1979. (Third edition).

Hutton, Charles. *A Philosophical and Mathematical Dictionary*. London: Charles Hutton, 1815.

Janes, W. C. "The Duodecimal System," *The Mathematics Teacher*, 37:365–367, December, 1944.

Johnson, J. T. "New Number Systems vs the Decimal System," *School Science and Mathematics*, 40:828–834, December, 1940.

Johnson, William Woolsey. "Octonary Numeration," *Bulletin of the New York Mathematical Society*, 1:1–6, 1891.

Johnston, James Halcro. *The Reverse Notation—Introducing Negative Digits with Twelve as Base*. London: Blackie and Son, 1938.

Jones, Phillip S. "Historically Speaking,—," *The Mathematics Teacher*, 46:575–577, December, 1953.

______. "Numeration or Scales of Notation." Unpublished study guide for a course at University of Michigan, Ann Arbor, c. 1964.

______. "Notes on Numeration: Arithmetic on a Checkerboard, Numerals for the Blind," *School Science and Mathematics*, 78:481–488, October, 1978.

Jordaine, Joshua. *Duodecimal Arithmetick*. London: Robert Midgeley, 1687.

Juskevic, A. P. and E. Winter. *Leonhard Euler und Christian Goldbach*. Berlin: Akademie-Verlag, 1965.

Kahn, David. *The Codebreakers*. New York: Macmillan, 1967.

Karst, Edgar. "Algorithms that Use Two Number Systems Simultaneously," *Mathematics Magazine*, 40:91–97, March, 1967.

Kasner, Edward and James Newman. *Mathematics and the Imagination*. New York: Simon and Schuster, 1958.

Kempner, A. J. "Anormal Systems of Numeration," *American Mathematical Monthly*. 43:610–617, December, 1936.

Kiesswetter, Karl. "Ein einfaches Beispiel für eine Funktion, welche überall stetig und nicht differenzierbar ist," *Semesterberichte*, 13:216–221, 1966.

Klir, Jiri. "Weight Codes," *Stroje na spracovani informaci, sbornik*, 8:155–162, 1962.

Knuth, Donald E. "An Imaginary Number System," *Commu-*

nications of the Association for Computing Machinery, 3:245–247, 1960.

______. *Seminumerical Algorithms* (2nd edition)(vol. 2 of *The Art of Computer Programming*). Reading, Massachusetts: Addison-Wesley Publishing Co., 1981.

Köhlern, Heinrich. *Lehr-Saetze des Gottfried Wilhelm von Leibniz*. Leipzig: Joh. Meyers, 1720.

Kokomoor, Franklin Wesley. *Mathematics in Human Affairs*. New York: Prentice-Hall, 1942.

Lagrange, Joseph Louis. *Lectures on Elementary Mathematics*. Trans. Thomas J. McCormack. Chicago: The Open Court Publishing Co., 1898.

Lamarck, Jean Baptiste. *Flore française*. 3 vols. Paris: H. Agasse, 1778.

Landau, Edmund. *Vorlesungen über Zahlentheorie—Aus der elementaren Zahlentheorie*. New York: Chelsea Publishing Co., 1950.

Lange, Luise. "On Fingerprints in Number Words," *School Science and Mathematics*, 36:13–19, January, 1936.

Laplace, Pierre-Simon de. "Lecons de mathématiques données à l'école normale, en 1795." *Journal de l'école polytechnique*, VIe et VIIe Cahiers, 1812.

Larrivee, Jules A. "A History of Computers, II," *The Mathematics Teacher*, 51:541–544, November, 1958.

Larsen, Harold. "A Note on Scales of Notation," *American Mathematical Monthly*, 51:274–275, May, 1944.

Legendre, Adrien-Marie. *Essai sur la théorie des nombres*. Paris: Duprat, 1798.

Leibniz, Gottfried Wilhelm. "Explication de l'arithmétique binaire," *Memoires de l'Academie Royale*, 1703:85–89, 1703.

______. *Opera Omnia*. 6 vols. Geneva: Fratres des Tournes, 1768.

Leslie, John. *The Philosophy of Arithmetic*. Edinburgh: Archibald Constable and Company, 1817.

Lucas, Edouard. *Récréations Mathématiques*. 4 vols. Paris: Gauthier-Villars et Fils, 1891.

———. *Théorie des Nombres*. Paris: Gauthiers-Villars et Fils, 1891.

McCracken, Daniel D. *A Guide to FORTRAN Programming.* New York: Wiley, 1961.

———. *A Guide to FORTRAN IV Programming*. New York: Wiley, 1965.

McIntire, D. P. "A New System for Playing the Game of Nim," *American Mathematical Monthly*, 49:44–46, January, 1942.

McLeod, Herbert, ed. *Mathematics*. Vol. I of *The Subject Index of the Royal Society of London Catalogue of Scientific Papers*. 3 vols. Cambridge: Cambridge University Press, 1908–14.

Mackenzie, Charles E. *Coded Character Sets, History and Development*. Reading, Massachusetts: Addison-Wesley Publishing Co., 1980.

MacMahon, P. A. "Certain Special Partitions of Numbers," *Quarterly Journal of Pure and Applied Mathematics*, 21: 367–373, 1886.

———. "The Theory of Perfect Partitions of Numbers and the Compositions of Multipartite Numbers," *Messenger of Mathematics*, 20:101–119, 1891.

———. "Weighing by a Series of Weights," *Nature*, 43:113–114, 1891.

Mahnke, Dietrich. "Leibniz auf der Suche nach einer allgemeinen Primzahlgleichung," *Bibliotheca mathematica*, (3)13:29–61, 1912–13.

Mariage, Aimé. *Numération par huit*. Paris: Le Normant, 1857.

Mirick, Gordon R. and Vera Sanford. "Scales of Notation," *The Mathematics Teacher*, 18:465–71, 1925.

Mirsky, L. "A Theorem on Representations of Integers in the Scale of r," *Scripta Mathematica*, 15:11–12, 1949.

Moore, E. H. "Nim," *Annals of Mathematics*, series 2, 11: 93–94, 1910.

Müller, Felix. "Über eine Zahlentheoretische Spielerei," *Zeitschrift für Mathematik und Physik*, 21:227–228, 1876.

National Council of Teachers of Mathematics. (NCTM), *The Revolution in School Mathematics*, Washington, D.C.: National Council of Teachers of Mathematics, 1961.

______. *Topics in Mathematics for Elementary School Teachers*. Washington, D.C.: National Council of Teachers of Mathematics, 1964.

Neugebauer, Otto. *The Exact Sciences in Antiquity*. Princeton: Princeton University Press, 1952.

Nolte, Rudolf August. *Leibniz Mathematischer Beweis der Erschaffung und Ordnung der Welt in einem Medallion an den Herrn Rudolf August*. Leipzig: J. C. Langenheim, 1734.

Norwood, James M. G. "I Ching," *Harper's Bazaar*, 101:219, October, 1968.

Nystrom, John William. *On the French Metric System: with a Discussion of a Duodecimal Notation*. Philadelphia, Pa.: J. Pennington and son, 1876.

______. *Project of a New System of Arithmetic, Weight, Measure and Coins, Proposed to be Called the Tonal System with Sixteen to the Base*. Philadelphia: J. B. Lippincott, 1862.

Olmstead, John M. H. *The Real Number System*. New York: Appleton-Century-Crofts, 1962.

Ore, Oystein. *Invitation to Number Theory*. New York: Random House, 1967.

Ozanam, Jacques. *Récréations mathématiques et physiques*. Paris: Jean Jombert, 1698.

______. *Récréations mathématiques et physiques, nouvelle edition, revue, correige et augmentée*. 4 vols. Paris: Charles Antoine Jombert, 1735–36.

______. *Recreations in Mathematics and Natural Philosophy, enlarged by Montucla*. Trans. Charles Hutton. 4 vols. London: G. Kearsley, 1803.

______. *Recreations in Mathematics and Natural Philosophy, enlarged by Montucla*. Trans. Charles Hutton. 4 vols. London: Longman, Hurst, Rees, Orme, and Brown, 1814.

Pascal, Blaise. *Oeuvres*. 14 vols. Paris: Brunschvicg et Boutroux, 1908.

Pawlak, Z. and A. Wakulicz. "Use of Expansions with a Negative Basis in the Arithmometer of a Digital Computer," *Bulletin de l'academie polonaise des sciences*, 5:233–236, March, 1957.

Peano, Giuseppe. "La numerazione binaria applicata alla stenografia," *Atti della R. Academia delle Scienze di Torino*. 34:47–55, 1899.

______. *Formulaire de mathématiques*. Tome 3. Paris: Gauthier-Villars et Fils, 1901.

Pedoe, Dan. *The Gentle Art of Mathematics*. New York: The Macmillan Co., 1958.

Penney, Walter, "A 'Binary' System for Complex Numbers," *Journal of the Association for Computing Machinery*, 12:247–248, April, 1965.

Peterson, W. Wesley, *Error-correcting Codes*. Cambridge, Mass.: M.I.T. Press, 1961.

Philastre, P. L. F. "Le Yi: King our Livre des changements de la dynasti des Tsheou," *Annales du Musee Guimet*, 8:1–489, 1885 and 23:1–608, 1893.

Phillips, E. William. "Binary Calculation," *Journal of the Institute of Actuaries*, 67:187–221, 1936.

Pierce, Robert M. *Problems of Number and Measure*. Chicago: Robert M. Pierce, 1898.

Pitcher, Wilimina E. "Alice in Dozenland," *The Mathematics Teacher*, 27:390–396, December, 1934.

Pitman, Benn. *Sir Isaac Pitman*. Cincinnati: C. J. Krehbiel Co., 1902.

Pitman, Isaac. "A New and Improved System of Numeration and Measurement," *The Phonetic Journal*, 66–69, February 9, 1856.

Posey, L. R. "Change of Base," *School Science and Mathematics*, 46:871–78, December, 1946.

"Proposed Duodecimal System," *Scientific American*, 87:440, December, 1902.

Pujals de la Bastida, D. Vicente. *Filosofia de la Numeración*. Barcelona: Los Herederos de la viuda Pla, 1844.

Recht, L. S. "The Game of Nim," *American Mathematical Monthly*, 50–435, August-September, 1943.
Redheffer, Raymond. "A Machine for Playing the Game of Nim," *American Mathematical Monthly*, 55:343–349, June-July, 1948.
Richards, R. K. *Arithmetic Operations in Digital Computers*. Princeton, New Jersey: D. Van Nostrand Co., 1955.
______. *Electronic Digital Systems*. New York: Wiley, 1966.
Rose, Kenneth. "Tradition vs. Octonary Arithmetic," *The New Philosophy*, 60:142–147, January, 1957.

Sawyer, W. W. *Vision in Elementary Mathematics*. Baltimore: Penguin Books, 1964.
Schaaf, William L. "Scales of Notation," *The Mathematics Teacher*, 47:415–417, October, 1954.
Scheffler, Hermann. *Beiträge zur Zahlentheorie*. Leipzig: Friedrich Foerster, 1891.
School Mathematics Study Group. *Experimental Units for Grades Seven and Eight*. New Haven: Yale University, 1959.
______. *Mathematics for Junior High School*. Vol. I, Part 1. New Haven: Yale University, 1961.
Seelbach, Lewis Carl. "Duodecimal Bibliography," *Duodecimal Bulletin*, 8:1–65, October, 1952.
Ser, J. *La Numération et le Calcul des Nombres*. Paris: Gauthier-Villars et Fils, 1944.
Shanks, Daniel. *Solved and Unsolved Problems in Number Theory*. (2nd edition) Washington, D.C.: Spartan Books, 1978.
Shannon, C. E. "A Symmetrical Notation for Numbers," American Mathematical Monthly, 57:90–93, 1950.
Shirley, John W. "Binary Numeration Before Leibniz," *American Journal of Physics*, 19:452–454, November, 1951.
______ (ed). *Thomas Hariot: Renaissance Scientist*. Oxford: Clarendon Press, 1974.
Simony, Oskar. "Über den Zusammenhang gewisser topologischer Thatsachen mit neuen Sätzen der höheren Arithmetik und dessen theoretische Bedeutung," *Sitzungs-*

berichte der Mathematisch-Naturwissenschaftlichen Classe der Kaiserlichen Akademie der Wissenschaften, Wien, 96 (II):191–286, 1887.

Smith, Cedric A. B. "Compound Games with Counters," *Journal of Recreational Mathematics*, 1:67–77, April, 1968.

Songster, G. F. "Negative-Base Number-Representation Systems," *IEEE Transactions on Electronic Computers*, (EC-12):274–277, June, 1963.

Sonnenschein, A. and H. A. Nesbitt. *The Science and Art of Arithmetic. Exercise Book. Part I.* London: W. Swan Sonnenschein and Allen, c. 1870.

______. *The Science and Art of Arithmetic. Part I.* 4th ed. London: Whittaker and Co., 1879.

______. *The Science and Art of Arithmetic. Part II and III.* 3rd ed. London: Whittaker and Co., 1877.

"Space Exploration," *Time*, 86:35–44, July 23, 1965.

Spencer, Herbert, "Against the Metric System," *Appleton's Popular Science Monthly*, 186–202, June 1896.

Sperry Rand Corporation. *Preliminary Hardware Reference Manual for 9200/9300 Systems*. New York: Sperry Rand Corporation, 1966.

Sprague, R. "Über mathematische Kampfspiele," *Tohoku Mathematical Journal*, 41:438–444, 1934–36.

Stahlberger, E. "Über einen Gewichtssatz, dessen Gewichte nach Potenzen von 3 geordnet sind," *Repertorium für Experimental Physik, für Physikalische Technik, Mathematische und Astronomische Instrumentenkunde*, 5:10–13, 1869.

Stein, Heinrich Wilhelm. "Über die Vergleichung der verschiedenen Numerations-systeme," *Journal für die reine und angewandte Mathematik*, 1:369–371, 1826.

Stein, Sherman K. *Mathematics, The Man-made Universe*. San Francisco: W. H. Freeman, 1963.

Stevin, Simon. *Les oeuvres mathématique*. Vol. I. Leyden: B. and A. Elsevier, 1618.

Stibitz, George R. "U.S. Patent No. 2, 486, 809—Biquinary System Calculator," November 1, 1949.

Stifler, W. W. ed. *High-speed Computing Devices*. New York: McGraw-Hill, 1950.

Struik, D. J. "Simon Stevin and the Decimal Fractions," *The Mathematics Teacher*, 52:474–479, October, 1959.

Studnicka, F. J. "Über eine neue Eigenschaft von Zahlen in 2n-zifferigen Systemen," *Sitzungsberichte der Königlichen Bömischen Gesellschaft der Wissenschaften*, VII: 7–10, 1896.

Swedberg, Emanual. *A New System of Reckoning Which Turns at 8*. Philadelphia: Swedenborg Scientific Association, 1941.

Tanner, H. W. Lloyd. "An Arithmetical Theorem," *Messenger of Mathematics*, 7:63–64, 1878.

Taylor, Alfred. "Octonary Numeration," *Proceedings of the American Philosophical Society*, 24:296–366, 1887.

Tennant, John. "On the Factorisation of High Numbers," *Quarterly Journal of Pure and Applied Mathematics*, 32:322–342, 1901.

Tenzeln, Wilhelm Ernst. *Curieuse Bibliothec*. Franckfurt: Philipp Wilhelm Stock, 1705.

Terry, George S. *Duodecimal Arithmetic*. London: Longmans Green and Co., 1938.

"The Duodecimal Society of America," *School Science and Mathematics*, 44:694, November, 1944.

Thiele, T. N. "Quel nombre serait à préférer comme base de notre système de numération?" *Danske Videnskarbenes Selskabs, Oversigt over det*, 1889:25–42, 1889.

Tingley, E. M. "Calculate by Eights, Not by Tens," *School Science and Mathematics*, 34:395–399, April, 1934.

Tropfke, Johannes. *Geschichte der Elementar-Mathematik*. Berlin: Walter de Gruyter und Co., 1937.

Unger, Friedrich. *Die Methodik der Praktischen Arithmetik*. Leipzig: B. G. Teubner, 1888.

Uspensky, J. V. and M. A. Heaslet. *Elementary Number Theory*. New York: McGraw-Hill, 1939.

Vellnagel, Christoph F. *Numerandi methodi*. Jena: Franciscus, 1740.

Vincent, L. H. "Duodecimal System of Notation," *School Science and Mathematics*, 9:555–62, June, 1909.

Voltaire, Francois Marie. *Histoire de Charles XII., Roi de Suede*. Basle: Christophe Revis, 1781.

______. *Oeuvres complètes*. Volume 22. Paris: P. Dupont, 1823.

Ware, Willis H. *Digital Computer Technology and Design*. New York: Wiley, 1963.

Watkins, Alfred. "Octaval Notation and the Measurement of Binary Inch Fractions," *American Machinist*. 52: 685–688, March, 1920.

Weeg, G. P. "Uniqueness of Weighted Code Representa-resentations," *IRE Transactions on Electronic Computers*, 9:487–489, December, 1960.

Weidler, Johann Friedrich. *Dissertatio mathematica de praestantia arithmeticae decadicae*. Wittenberg: Samuel Creusig, 1719.

Weigel, Erhard. *Philosophia mathematica*. Jena: Matth. Birckner, 1693.

______. *Tetractyn*. Jena: Johann Meyer, 1672.

Wiedeburg, Johann Bernard. *Dissertatio mathematica de praestantia arithmeticae binaria prae decimali*. Jena: Krebs, 1718.

______. *Einleitung zu den mathematischen Wissenschaften*. Jena: Joh. Meyers, 1725.

Wilkinson, James H. *Rounding Errors in Algebraic Processes*. Englewood Cliffs: Prentice-Hall, Inc., 1963.

William, R. P. "Ancient Duodecimal System," *School Science and Mathematics*, 9:516–21, June, 1909.

Wishard, G. W. "The Octo-binary System," *National Mathematics Magazine*, 11:253–54, March, 1937.

Wizel, Adam. "Ein Fall von phänomenalem Rechentalent bei einem Imbecillen," *Archiv für Psychiatrie und Nervenkrankheiten*, 38:122–155, 1904.

Wynn-Williams, C. E. "A Thyatron 'Scale of Two' Automatic Counter," *Proceedings of the Royal Society (London)*, 136:312–324, 1932.

Wythoff, W. A. "A Modification of the Game of Nim," *Nieuw Archief voor Wiskunde*, 7:199–202, 1907.

INDEX

Academie Royale des Sciences, 15, 44, 45, 46, 71, 187
Adams, John Quincy, 83
Ahrens, Wilhelm, 9, 118–19
Alder, H. L., 161
algorithms, 162–67
alphabet
  numeric values of, in binary, hexadecimal and decimal, 106 (fig. 11)
  See also codes "bi-literal"
American Philosophical Society, 108
American Society of Mechanical Engineers, 126
Ampère, André-Marie, 84–85, 86 (fig. 8), 104
Andrews, F. Emerson, 126–27
Archibald, Raymond Clark, 9, 51
"Astrognostich-heraldisches Collegium", 23
Atanasoff, Vincent, 138
Atanasoff-Berry computer, 137

Bachet, Claude-Gaspar, 15
Bachman, Paul, 118
Bacon, Francis, 15, 20, 169
Balasinski, W., 160
Ball, W. W. Rouse, 117, 119
Barlow, Peter, 15, 80–83, 85, 92–96 passim, 102, 118, 160, 171
base (2) *see* binary system
base (3), 15, 20, 62, 63, 82, 89–93 passim, 170
base (4), 20–24 passim, 28, 55, 62, 63, 90–91, 109–10, 121, 125, 162, 170, 173
base (5), 20, 62–65 passim, 71, 87, 170, 174
base (6), 20, 62, 63, 70, 121, 124, 170
base (7), 20, 62, 63, 166, 170, 174
base (8), 20, 57–60, 62, 63, 88, 101, 108–109, 120, 121–24 passim, 136, 170–73 passim
  conversion from base (2), 171
base (9), 20, 62, 63, 170
base (10) *see* decimal system
base (11), 72, 73
base (12), 9, 20, 24–28, 60–65 passim, 73, 79–82 passim, 87, 88, 110, 126–27, 169, 171–72, 173
  advantages of, 80, 87
  application to measurement, 28, 89, 124
  arithmetic operations of, 64
  conversion from decimal to, 73, 89
  fractions in, 44
  in metric systems, 71–72
  and Pascal, 25
base (13), 63
base (15), 63
base (16), 88, 99, 119, 122, 136, 171, 172
  numeric values of alphabet in, 106 (fig. 11)
  comparison of, with decimal system, 180 (table D)
base (24), 63
base (30), 63, 64, 69, 74
base (60), 20, 22, 65, 74, 170
base (64), 57, 60, 74, 170, 173
base (900), 119
base (β), 8, 21, 64
  Barlow and Cauchy modifications, 96
  conversion from base (10), 65
  divisibility rules of, 169
  fractions of, 66
  negative notation of, 160
  powers of, 21
bases, 173–74
  comparison of, 178 (table B)
  irrational *see* Bergman, Kempner
Bavarian Academy of Science, 65
Begle, E., 6
Beguelin, N., 61–62

Bell Telephone Company, 6, 147
 See also telephone
Bellavitis, G., 99–100
Bellman, R., 159
Berckenkamp, Johann, 63–64
Berdellé, Charles, 101, 173
Bergman, George, 160
Berkeley, Edmund, 138
Berlin Academy of Science, 39, 53
Bernoulli, Jacques, 46–47
Bernoulli, Johann (Jean), 39
Bezout, Etienne, 64, 66, 76n.10
binary arithmetic, 39–46 passim, 40 (fig. 5), 53, 55, 56 (fig. 7), 61, 65, 83, 98 (table 10), 99
binary coded decimals *see* decimal system, binary coded
binary fractions *see* fractions, binary
Binary Medallion, 31–34 passim, 51n.3, 55, 170
binary notation, 61, 101, 102
 columnar sums in, 116
 factors in, 102
 Pierce's 'improved', 98 (fig. 10)
 logarithms in, 55, 63
binary strings, 6, 15, 37, 61
 columnar periods in, 37, 46–47, 115, 116
 shrinkage of, 99
binary system, 3–5 passim, 13, 20, 62, 63, 81, 87, 99, 100, 135, 169, 170, 173
 conversion to base (8), 171
 discovery of, 13, 14, 44
 first publication on, 20
 medallion, *see* Binary Medallion
 modified, 96–97
 numeric values of alphabet in, 106 (fig. 11)
 navigational computation using, 113
 and other number representations, 177 (table A)
 similarity to base (8), 101
 and stenography, 104
 suitability for computers, 134–36, 172
 See also Bernoulli, binary arithmetic, binary notation, binary strings, Chow-tse's Diagram, recreational mathematics, Schulenberg, Wiedeburg
Biquinary Calculator, 139
Blij, F. von der, 49, 51
Bouton, Charles L., 116, 172
Bowden, Joseph, 120, 121
Brander, George, 65–66
Briggs, Henry, 45
Brockhaus Conversations-Lexicon, 100
Brooks, Edward, 110–11
Brown, J. L., 161
Bouvet, R. P., 42–43
Brocard, H., 51
Brunetti, Francesco, 62–63
Buchanan, Herbert, 128
Buffon, George, 64–65, 66
Burks-Goldstine-Von Neumann Report, 134–37, 172

Cajori, Florian, 23
Cantor, George, 102
Cantor, Moritz, 20, 23, 51, 53, 91, 93, 115
Caramuel y Lobkowitz, Bishop Juan, 14, 20–22, 25, 28, 46, 55, 74, 101, 115, 169, 170
Cardan's Rings *see* recreational mathematics
Carus, P., 50 (fig. 6), 51
Cauchy, Augustin-Louis, 96, 127, 128, 160
Charles XII of Sweden, 57–59, 60, 118, 173
Cheo, Peh-Hseuin, 159
China and binary system *see* recreational mathematics
Chow Tse's Diagram, 50 (fig. 6)
Christofferson, H. C., 126
Chrystal, George, 128, 129
codes
 alphanumeric, 151–53
 "bi-literal", 15
 binary, 169, 179 (table C)

binary vs decimal, 149–151
binary coded decimal, 6, 133, 179 (table C)
biquinary, 138, 139, 147
decimal, 139, 143–48 passim, 181 (table E)
8-4-2-1, 138, 139–43, 181 (table E)
ENIAC, 139, 147
error correcting, xv, 148, 149
excess three, 138, 139, 141, 143, 181 (table E)
four-bit, 139–45, 181 (table E)
hexadecimal, 181 (table E)
International Morse, 105
octal, 181 (table E)
reflected, or gray tones, 153–56
self-complementing, 140–43
weighted, 139, 142 (table 2)
Collignan, E., 96
columnar periods *see* binary strings, columnar periods in
combinatorics, 23
Commission on Mathematics
and binary system, 4
Committee on the Undergraduate Program in Mathematics (Berkeley, CA), 5
Complex Computer,138
computer applications, 133–58, 172
See also binary system
computer programming *see* FORTRAN
computers, electronic digital, 151
binary, 150 (table 4)
decimal, 133, 136, 150 (table 4)
pre-electronic, history of, 137–39
See also IBM 360 series computers, Burks-Goldstine-Von-Neumann Report, Complex Computer, ENIAC
Conant, Lev, 60
Condon, E. U., 117
Cooper, G. H., 120
Counters, electronic, 137
Courant, Richard, 125
Cova, Figures of the Eight *see* Fohy, Figures of
Cunningham, Allan, 120

Dangicourt, Petr, 53–54, 55, 61, 116
De Augmentis Scientarum, 15, 169
de Fermat, Pierre, 48
Conjecture, 61
decimal system, 20, 79
binary-coded, 133, 136, 179 (table C)
"casting out nines", 81, 87
codes, 138, 139
coinage in, 60
comparison with base (16), 180 (table D)
conversion to binary, 74
divisibility properties of, 18, 169
duodecimal vs decimalized metric system, 71–72
modification of, 85
numeric value of alphabet in, 106 (fig. 11)
decimalization
advocates of, *see* Stevin, Swedborg, Stein
DeMorgan, Augustus, 87–90 passim, 171
"De numeris multiplicibus", 15–19, 20
De Thiende, 14
Descartes, René, 23
Dickson, Leonard, 9, 69, 100
Disquisitiones Arithmeticae, 79, 100
divisibility property of numbers, 16–18, 19 (table 1), 22, 169–72 passim and Pascal's theorem, 72–73, 79, 81
Drazin, M. P., 159
Duke of Brunswick *see* Rudolph August, Duke of Brunswick
Duodecimal Arithmetick, 24
Duodecimal Society of America, 127
duodecimal system *see* base (12)

DuPasquier, Louis-Gustave, 121
dyadics *see* binary system

Eccles-Jordan flip-flop trigger circuit, 134, 137
Eggan, L. C., 162
ENIAC, 133, 136, 139, 140, 147, 172
Euler, Leonard, 61, 75, 96
"Explication de l'arithmetique binaire", 20, 34, 35, 39–43, 170
  arithmetic operations in, 40 (fig. 5)
  binary interpretation of Figures of Fohy, 39, 41–43, 49
  sequels to, 53–55
  Table of Numbers in, 38 (fig. 4)
exponents, 53, 75–76
  in β systems, 65

fedebra, 53
Felkel, Anton, 66–71, 74, 79
Fialkowsky, Konrad, 161
Fine, N. J., 161
Fohy, Figures of, 39, 41–43, 79, 101
  reactions to Leibniz interpretation of, 49–51, 53
Fontenelle, Bernard Le Bovier de, 44, 45, 55, 187
FORTRAN, 4, 151
Foss, F. A., 157
fractions
  β-adic, 66–71
  and base (12), 44
  "Binal", 120
  binary, 62–53, 151
  binary and decimal, 48, 49
  conversion from decimal to base (12), 27, 28
  decimal and binary equivalents of, 169, 182–85 (tables F through I)
  decimal and nondecimal, 28, 79, 83
Fraenkel, Abraham, 124
French Metric Commission, 173

Games *see* recreational mathematics
Gauss, Carl, 79, 81, 100
Gelin, E., 109, 173
Gilbert, E. N., 155
Glaisher, J. W. L., 90
Goldbach, Christian, 61
Goldstine, Herman and Adele, 140
  See also Burks-Goldstine, Von Neumann Report, 134–37, 172
Gray, F., 157
Griffith, J. Stanley, 159
Grossman, H. D., 117
Grünwald, Vittorio, 97
Guhrauer, G. E., 23

Hamming, R. W., 149
Hankel, Herman, 90–91, 173
Hardy, G. H., 117, 118–19
Hariot, Thomas, 11–14, 15, 20, 22, 41, 169, 170
Harkin, Duncan, 125
Heath, Thomas, 23
hexadecimal system *see* base (16)
Hierocles, 23
Hindu-Arabic system, 5–8 passim, 169
Histoire de l'Academie Royale des Sciences, 44, 45, 55, 187
Hocevar, Franz, 100, 101
Hutton, Charles, 79, 83

I Ching, 49, 51
IBM 360 series computers, 153
Institute for Advanced Study (Princeton, NJ), 136
International Decimal Association, 89
International Metric System, 171
International Morse Code *see* codes, International Morse

Janes, L. C., 127
Johnson, J. T., 124
Johnson, William Woolsey, 108–09

Johnston, James, 127
Jones, Philip S., 3
Jordaine, Joshua, 24–28, 69, 170

Kahn, David, 15
Karst, Edgar, 162
Kempner, A. J., 121–22, 23
Kieswetter, Karl, 162
Klir, Jiri, 161
Knuth, Donald E., 160, 167
Köhlern, Heinrich, 35
Kokomoor, Franklin, 128–29

Lagny, Thomas Fantet de, 44–46 passim, 55, 125, 170, 173, 187
Lagrange, Joseph, 72–73, 173
Lamarck, Jean
  and binary system, 73, 104
Lambert, Johann Heinrich, 48
Landau, Edmund, 119
Lange, Luise, 127
LaPlace, Pierre-Simon, 73
Larsen, Harold, 125
Legendre, Adrien-Marie, 74–76, 100, 101, 124, 128
"Legendre" symbol, 75
Leibniz, Gottfried Wilhelm, 14, 15, 20, 23, 29, 44, 31–52, 53–55 passim, 65, 73, 79, 83, 85, 99, 101, 115–16, 170, 173
  "Explication", 39–43
  letter to Schulenberg, 36 (fig. 3), 37
  missionary endeavors of, 31, 33, 73
  See also Binary Medallion
Leslie, John, 83–84
Lobkowitz, Bishop Juan Caramuel y *see* Caramuel Lobkowitz, Bishop Juan
logarithms, 21, 45–46, 126,
  in binary notation, 55, 63
Lucas, Edward, 102, 118

McCracken, Daniel, 4, 151
McIntyre, D. P., 117
McLeod, Herbert, 8
MacMahon, P. A., 93–96
Mahnke, Dietrich, 30, 47–49, 79
Mariage, Aimé, 88
Mariner IV, 5, 172
Mars
  photographs of, 5, 172
Mathematical Institute (Utrecht, Netherlands), 51
Mathematics Society of Hamburg, 35
Mathematics Tribunal, 33
Mathesis biceps see "Meditatio"
Mauchley, John W., 133, 138
  See also ENIAC
measurement *see* weights and measurement
"Meditatio," 20–22, 23
Memoires de l'Academie Royale des Sciences, 39, 44, 53, 187
Metius, Peter, 83
Metric Commission, 71–72, 73
metric system, 71–72, 83, 89, 124, 172, 173
  decimalized, 108, 110, 171–72
Mirsky, L., 159
modular congruences, 79, 81
monadology, 35
Moore, E. H., 117
Mrowka, S., 160
Müller, Felix, 91–93, 95, 96
multiplication
  nondecimal, 22, 40 (fig. 5), 41
  Russian peasant *see* recreational mathematics
musical scales, 20, 21, 23

Napier's Bones, 60
National Council of Teachers of Mathematics
  and nondecimal numeration, 4
Nim, Game of, 10, 116–17, 118, 119, 125, 165
  modification of, 117, 165, 172
  Nimatron, 117
Nolte, Rudolph August, 32 (fig. 1), 34 (fig. 2), 35
nondecimal numeration, 20, 79, 80–83, 87, 90, 164 (table 7), 171

biblical sources of, 166
and computers, 3, 4, 5
contemporary literature on, 159–63
contemporary reference books about, 165–67
parent involvement in, 6, 7
and pedagogy *see* teaching of nondecimal numeration
textbooks on, 127–129
Nordberg, Goran, 57
Norwood, James, 49
notation
γ-adic coded β-adic base, 74–75
of negative base β, 160
binary see binary notation
"Nouvelle Arithmetique", 44, 45, 55, 187–92
numbers
formula of prime, 48
irrational, 48–49
natural, 45
properties of, 44
numeration systems
bibliographies and study guide to, 9
comparisons of, 84, 178 (table B), 180 (table D)
development of, 169–72
first scientific treatment of, 15
historical surveys of, 9, 87
miscellaneous publications on, 119–26
nondecimal *see* nondecimal numeration
nonstandard, 7, 125–26, 161
"Quater-imaginary", 160–61
role of, in recreational mathematics and theory of numbers, 118–19
standard, 7, 8, 25
See also bases (3, 4, 5, 6, 7, 8, 9, 11, 12, 13, 15, 16, 24, 30, 60, 64, 900, β)
Nystrom, John William, 88, 172

octal system *see* base (8)
Ozanam, Jacques, 79–80, 93

Pascal, Blaise, 15–19, 20, 22, 25, 26, 28, 169
Pascal's General Divisibility Theorem, 16, 17, 19 (table 1), 72–73, 79, 81
Peano, Giuseppe, 85, 104–05, 106 (fig. 11), 107, 153
Pedoe, Don, 119
Peirce, Benjamin, 99
Peirce, Charles, 99
Penney, Walter, 162
Peterson, W. Wesley, 149
Philastre, P. L. F., 51
Phillips, E. William, 121
Philosophia Mathematica, 28
Pierce, Robert, 103
Pitcher, Wilimina, 127
Pitman, Sir Isaac, 88, 171
Posey, L. R., 126
Pujals dc la Bastida, D. Vicente, 87
Pythagoreans, 21, 23 passim

quaternary system *see* base (4)

Recht, L. S., 117
recreational mathematics, 65, 79, 92–93, 102, 115–19
Cardan's Rings, 10
Chinese Rings, 118
Chow Tse's Diagram, 50 (fig. 6)
Fohy, Figures of, 39, 41–43, 49–51, 53, 79, 100
Game of Nim, *see* Nim, Game of
"Guess the Number Cards", 93
I Ching, 49, 51
Muller Cards, 92, 94 (fig. 9)
Russian Peasant Multiplication, 10, 120, 125
"Tell Your Age" game, 91
Tower of Hanoi, 118
Weighing Problems, 95, 125
Redheffer, Raymond, 117
Richards, R. K., 136, 137, 143, 145
Roman numerals, 5
Rose, Kenneth, 159–60

Rudolph August, Duke of Brunswick, 31–35 passim, 37, 39, 170

Sawyer, W. W., 3
Schaaf, William, 8
Scheffler, Hermann, 102
School Mathematics Study Group (SMSG)
  and nondecimal numeration, 4, 165
Schulenberg, Johann Christian, 36 (fig. 3), 37
"secundals", 99
Seelbach, Lewis, 8, 9
Ser, J., 125–26
sets and subsets, 11, 13, 20
  See also musical scales
sexagesimal system *see* base(6)
Shannon, C. E., 160
Shapiro, H. N., 159
Shirley, J. W., 14, 169
Simony, Oskar, 101
Smith, Cedric, 117
Societas Pythagorea, 23
Songster, G. F., 162
Sonnenschein, A., 90, 111
space exploration *see* Mars, Mariner IV
Spencer, Herbert, 110
Sprague, R., 117
Stahlberger, E., 89–90
Stein, Heinrich Wilhelm, 84, 90, 173
stenography, 104
Stevin, Simon, 14, 15, 80, 102, 118, 169
Stibitz, George R., 138
Struik, D. J., 14
Studnicka, F. J., 103
Swedborg *see* Swedenborg, Emanuel
Swedenborg, Emanuel, 57–60, 88, 160, 170–73 passim

Tanner, Lloyd, 100
Taylor, Alfred B., 108, 123–24
teaching of nondecimal numeration, 3–5 passim, 8, 9, 72, 73, 109–10, 111, 162–64, 174
telephone
  and binary digit strings, 6
  switching systems in, 137
  codes in, 146, 147
Tenant, John, 119
Tentzeln, Wilhelm Ernest, 49
Terry, George, 127
Tetractyn, 22–24 passim
theorems
  β-adic fractions, 67–69
  γ-adic coded β-adic string, 74–75
  Berckenkamp Theorem (57), 63–64
  canonical numbers, 122–23
  codes and weighting, 145
  divisibility, 15, 16, 28, 72
  Fermat's "Last Theorem", 48
  Fine and Conjecture, 161–62
  Glaisher, J. W. L., 90
  Hardy (decimals), 119
  natural numbers, 13, 41
  negative bases, 160
  rational base expansion, 162
  remainders in nondecimal numeration, 81
  sets and subsets, 13
Thiele, T. N., 109–10, 173
Tingley, E. M., 121, 124
trigonometry, 45
Tropfke, Johannes, 15, 20

Unger, Friedrich, 7
Uspensky, J. V., 117, 119

Vanden Eynden, C. L., 162
Vellnagel, Christoph Friedrich, 62
Vincent, L. H., 126
Voltaire, 57

Wakulicz, A., 160
Ware, Willis, 143
Watkins, Alfred, 120
Weeg, G. P., 145, 161
Weidler, Johann Friedrich, 55
Weigel, Erhard, 22–24, 28, 55,

115, 170
weights and measurement, 28, 92, 93, 119, 127, 170–73 passim
Bachet's Problem of the Weights, 15
and base (2), 15, 41, 55, 81–82
and base (16), 88, 172
decimalizing of, 14, 60, 80, 83, 171, 173
"either pan" weighing problem, 89
in metric system *see* metric system
"weighing problem" letters, 95
Wiedeburg, Johann Bernard, 32 (fig. 1), 55, 56 (fig. 7), 65
William, R. P., 126
Wishard, G. W., 123–24
Wizel, Adam, 119
Wollis, John, 48
Wynn-Williams, C. E., 137
Wytoff, W. A., 117

Yien, Sze-Chien, 159